# BLOOD LIPIDS AND LIPOPROTEINS

# BIOCHEMISTRY, DISORDERS AND ROLE OF PHYSICAL ACTIVITY

# PROTEIN BIOCHEMISTRY, SYNTHESIS, STRUCTURE AND CELLULAR FUNCTION

Additional books in this series can be found on Nova's website
under the Series tab.

Additional e-books in this series can be found on Nova's website
under the e-book tab.

PROTEIN BIOCHEMISTRY, SYNTHESIS, STRUCTURE AND CELLULAR FUNCTION

# BLOOD LIPIDS AND LIPOPROTEINS

## BIOCHEMISTRY, DISORDERS AND ROLE OF PHYSICAL ACTIVITY

MELISSA R. RUIZ
EDITOR

New York

Copyright © 2015 by Nova Science Publishers, Inc.

**NOTICE TO THE READER**

**Library of Congress Cataloging-in-Publication Data**

ISBN: 978-1-63482-591-7
Library of Congress Control Number: 2015936700

*Published by Nova Science Publishers, Inc. † New York*

# CONTENTS

**Preface**                                                                    **vii**

**Chapter 1**     Exercise and the HDL Quality                                  **1**
                  *Kazuhiko Kotani*

**Chapter 2**     Naturally Occurring Multiple-Modified
                  Low Density Lipoprotein                                       **13**
                  *Alexander N. Orekhov, Ekaterina A. Ivanova*
                  *and Yuri V. Bobryshev*

**Chapter 3**     Regulation of the Intake of Arachidonic Acid by
                  Modifying Animal Products and Its Effect on
                  Inflammatory Processes in the Human Body                      **55**
                  *Dorota Bederska-Łojewska, Marek Pieszka,*
                  *Paulina Szczurek, Sylwia Orczewska-Dudek*
                  *and Mariusz Pietras*

**Index**                                                                      **81**

# PREFACE

Circulating high-density lipoprotein (HDL) cholesterol (HDL-C) is a marker associated with cardiovascular health. Exercise is generally known to increase the HDL-C levels, and this can, in part, explain its cardioprotective effects. The authors present data regarding the association between exercise and the HDL quality and further encourages taking into consideration the view of HDL quality in relation to exercise, in addition to HDL-C. In contrast, this book also includes research on low density lipoproteins, specifically describing research in which atherogenic LDL possesses numerous alterations of carbohydrate and human blood plasma and represents a cascade of successive changes in the lipoprotein particle. Lastly, arachidonic acid has shown to effect blood lipid levels. The authors of this book focused on the problem of arachidonic acid metabolism, belonging to the group of n-6.

Chapter 1 - While HDL-C is a marker of the HDL quantity, special attention to the HDL quality (reflective of HDL functionality) has recently arisen as a new view on HDL biology and clinical studies of HDL. However, there is less information on the changes in the HDL quality induced by exercise. When the authors present some data regarding the association between exercise and the HDL quality, i.e., using oxidatively modified HDL and cholesterol efflux functionality of HDL, exercise may favorably affect the HDL quality. The present paper further encourages taking into consideration the view of HDL quality in relation to exercise, in addition to HDL-C.

Chapter 2 - Extra- and intracellular deposition of lipids, predominantly of cholesteryl esters, in arterial intima is one of the earliest manifestations of atherosclerosis. Formation of lipid-laden foam cells is recognized as a trigger in the pathogenesis of atherosclerosis. Low-density lipoprotein (LDL) circulating in human blood is the source of lipids accumulated in arterial cells.

However, numerous attempts to induce intracellular accumulation of cholesteryl esters by native LDL failed. On the other hand, LDL chemically modified in vitro (acetylated, malondialdehyde-treated, oxidized with ions of transient metals, etc.) induced lipid deposition in cells, i.e., is atherogenic. At the same time, the attempts to find LDL in circulation, similar to in vitro modified LDL, were unsuccessful. Thus, a paradoxical situation existed: 1) LDL is the source of lipids accumulated in the vascular wall; 2) native LDL does not induce accumulation of intracellular fat (foam cell formation); 3) in vitro modified forms of LDL are atherogenic, but nobody found them in circulation. In response to this challenge, intensive search for modified forms of circulating LDL have been started, and such forms were soon found. Among modified forms of LDL detected in blood, such forms as oxidized, small dense, and electronegative have been described. The authors discovered modified desialylated LDL in human blood plasma. This LDL was capable of inducing lipid accumulation in cultured cells. The authors isolated atherogenic LDL and characterized it. It was shown that atherogenic LDL possesses numerous alterations of carbohydrate, protein, and lipid moieties, and can be termed multiple-modified LDL. Multiple modification of LDL occurs in human blood plasma and represents a cascade of successive changes in the lipoprotein particle: desialylation, loss of lipids, reduction in the particle size, increase of surface electronegative charge, etc. In addition to intracellular lipid accumulation, stimulatory effects of naturally occurring multiple-modified LDL on other processes involved in the development of atherosclerotic lesions, namely cell proliferation and fibrosis, were shown.

Chapter 3 - In this study the authors focused on the problem of arachidonic acid metabolism, belonging to the group of *n-6*. In recent years, there has been a marked increase in human consumption of polyunsaturated fatty acids *n-6*, with a simultaneous reduction of *n-3* intake. Arachidonic acid, a component of the lipid bilayer, is a precursor of many biologically important compounds include eicosanoids (prostaglandins, leukotrienes) involved in the stimulation of inflammatory processes. To prevent this, our diet should contain foods rich in *n-3*, which compete with *n-6* for the same metabolic pathways, thereby reducing the level of arachidonic acid in cells and extinguish the inflammation. This can be achieved by changing diet and modifying the composition of products of animal origin, such as eggs, milk and meat. Studies on the properties of *n-3* and *n-6* have been widely carried out in terms of their pharmacological use in the treatment of diseases with acute and chronic inflammation.

In: Blood Lipids and Lipoproteins
Editor: Melissa R. Ruiz

ISBN: 978-1-63482-591-7
© 2015 Nova Science Publishers, Inc.

*Chapter 1*

# EXERCISE AND THE HDL QUALITY

## *Kazuhiko Kotani**

Department of Clinical Laboratory Medicine,
Jichi Medical University, Tochigi, Japan

## ABSTRACT

Circulating high-density lipoprotein (HDL) cholesterol (HDL-C) is a marker associated with cardiovascular health. Exercise is generally known to increase the HDL-C levels, and this can, in part, explain its cardioprotective effects. While HDL-C is a marker of the HDL quantity, special attention to the HDL quality (reflective of HDL functionality) has recently arisen as a new view on HDL biology and clinical studies of HDL. However, there is less information on the changes in the HDL quality induced by exercise. When we present some data regarding the association between exercise and the HDL quality, i.e., using oxidatively modified HDL and cholesterol efflux functionality of HDL, exercise may favorably affect the HDL quality. The present paper further encourages taking into consideration the view of HDL quality in relation to exercise, in addition to HDL-C.

* Corresponding author: Kazuhiko Kotani, PhD, MD, Department of Clinical Laboratory Medicine, Jichi Medical University, 3311-1 Yakushiji, Shimotsuke-City, Tochigi, 329-0498, Japan, Tel: +81-285-58-7386, Fax: +81-285-44-9947, E-mail: kazukotani@jichi.ac.jp.

# ASSOCIATION OF HIGH-DENSITY LIPOPROTEIN WITH CARDIOVASCULAR HEALTH

Lipid and lipoprotein disorders are relevant pathologic entities associated with cardiovascular health [1, 2]. Circulating high-density lipoprotein (HDL) cholesterol (HDL-C) is a well-known marker of cardioprotection; that is, relatively high HDL-C levels are inversely related to cardiovascular disease [3, 4]. Although a deeper understanding of HDL is still required, cardioprotection, as mediated by HDL, is explained by the following mechanisms: HDL exerts the efflux function of cholesterol from lipid-laden macrophages in the arterial wall and then delivers cholesterol to the liver, termed the reverse cholesterol transport pathway [5]. Cellular transporter molecules, such as adenosine triphosphate-binding cassette transporter A1, are involved in the step of cholesterol efflux achieved by HDL [5]. HDL inhibits the oxidation of atherogenic lipoproteins, such as low-density lipoprotein (LDL) [6] and exerts anti-inflammatory effects [7]. Various HDL-mediated mechanisms, including the inhibition of glycation, homocysteinylation, apoptosis and thrombosis, as well as the enhancement of nitric oxide, have also been documented [8-10]. These functions are, in part, due to the fact that HDL includes many protein molecules [7, 11].

# ASSOCIATION OF EXERCISE WITH HDL-C

A habit of exercise, one of physical activities, is considered to favorably affect the development of lipid and lipoprotein disorders, and the effects of exercise on lipid and lipoprotein metabolism are best investigated by focusing on the HDL-C levels in particular [12, 13]. Well-summarized reviews have described that the HDL-C level increases 4-18% by exercise, while the LDL-cholesterol level is not changed by exercise [12]. Several lines of evidence follow these observations, such as the association of exercise with an increased HDL-C level, regardless of an older age, gender and various disease conditions [14-16]. Even though conflicting data exist, this phenomenon is thought to be due to the different characteristics and methodologies of exercise used across studies (i.e., intensity, amount, frequency, duration [acute/chronic/intermittent] and type [aerobic/resistance]) [12]. Thus, the general consensus is that exercise mildly-to-moderately but favorably moderates HDL metabolism, helping to promote cardiovascular health.

Exercise is currently recognized to be a cornerstone modality for improving cardiovascular outcomes.

HDL comprises a heterogeneous family of particles, differing in density, size, membrane charge and lipid and protein composition [11, 17]. Based on the subfraction (subclass/subpopulation) of HDL particles, the HDL-C level is basically divided into the HDL2-C (large-sized) and HDL3-C (small-sized) levels [17]. Evidence indicates that exercise increases the HDL2-C level [18]. Recent studies also highlight the finding of increased HDL2-C levels induced by exercise [19-21], possibly contributing to increased total HDL-C levels as a result of exercise.

## PARADIGM SHIFT ON THE VIEW OF HDL

While the effects of the HDL subfraction on cardiovascular outcomes remain to be established [22], new insights have been obtained regarding the different roles of the subfraction of HDL in terms of cardioprotective properties [11, 23]. Surely, low HDL-C levels predict the development of cardiovascular disease, although interventional studies have revealed that the HDL level, as elevated by drugs such as cholesteryl ester transfer protein inhibitors, does not always protect against poor cardiovascular outcomes [24, 25]. Therefore, the associations between HDL and cardiovascular risks are recognized to be more complex than previously thought [7, 17, 23]. The total HDL-C levels express the nearly overall number of HDL particles; however, we think that the 'quantity of HDL' is not necessarily a perfect marker with respect to the evaluation of HDL in cardiovascular practice. Therefore, the principle of 'HDL quality' has arisen as a new concept 'beyond HDL-C,' which is considered to rely on the functionality of HDL, rather than HDL-C [23, 25-27]. Nonetheless, which measurements and/or tests are suitable for evaluating HDL and application in cardiovascular practice remains to be determined [28].

## ASSOCIATION OF EXERCISE WITH OXIDIZED HDL AS A MARKER OF HDL QUALITY

Apolipoprotein A-I (apoA-I) is known to be modified at several specific residues (i.e., methionine and tyrosine residues) [29-31]. ApoA-I is thought to be anti-oxidative, but more susceptible to oxidation than proteins on LDL [32-35]. The modification of apoA-I may be associated with 'HDL quality,' including the cholesterol efflux function of HDL [36]. As apoA-I is the most major protein component of HDL, the measurement of oxidatively modified apoA-I may be a marker of oxidized HDL (oxHDL).

We recently developed an ELISA system to measure the oxHDL levels [37, 38]. This assay utilizes an antibody against oxidized human apoA-I generated by treatment with $H_2O_2$, and does not react with native HDL, but rather a broad range of oxidative substances of HDL [37]. Our studies using this assay have shown increased oxHDL levels in patients with prediabetes and diabetes, an oxidative condition [37, 38]. Since no studies have characterized the relationship between exercise and oxHDL, we conducted a pilot study (with a single arm and small sample design) to see the changes in the oxHDL levels among 11 males (mean age 66 years) during a 6-week exercise intervention (Kotani K., unpublished data). In this population, while the overall HDL-C levels were reduced (this finding appears to be somewhat unexpected considering the above general consensus on the changes in HDL-C induced by exercise), the oxHDL levels tended to be reduced by increased exercise (Table 1). Of note, during the exercise intervention period, the changes in the HDL-C levels correlated inversely ($r = -0.62$, $p < 0.05$) with those in the levels of oxHDL/HDL-C, an index of HDL quality. Therefore, oxidatively modified HDL might have been improved in individuals with increased/sustained HDL-C levels during the study period.

**Table 1. Changes of parameters at the pre- and post-intervention**

| Parameters | Pre | Post | P |
|---|---|---|---|
| Body mass index, kg/m$^2$ | 22.7 ± 1.6 | 22.9 ± 1.7 | NS |
| Glucose, mg/dL | 105 ± 11 | 100 ± 9 | 0.05 |
| Total cholesterol, mg/dL | 231 ± 40 | 226 ± 36 | NS |
| Triglyceride, mg/dL | 96 (66-111) | 96 (76-134) | NS |
| HDL-C, mg/dL | 66 ± 12 | 56 ± 8 | < 0.01 |
| OxHDL, U/mL | 219 ± 56 | 178 ± 47 | 0.09 |
| OxHDL/HDL-C | 3.4 ± 0.9 | 3.2 ± 1.0 | NS |

Data are expressed as mean ± standard deviation or median (interquartile range).
NS: no significance, HDL-C: high-density lipoprotein cholesterol, oxHDL: oxidized high-density lipoprotein.
P: paired t-test (pre and post). Triglyceride values were log-transformed.

Recently, an additional marker of oxHDL was developed by other investigators, and that study showed high levels of oxHDL in patients with cardiovascular disease [40]. Research on the association of oxHDL with cardiovascular outcomes in relation to exercise is ongoing.

# ASSOCIATION OF EXERCISE WITH CHOLESTEROL EFFLUX OF HDL AS A MARKER OF HDL QUALITY

Cholesterol efflux assays have been applied, and studies using these assays have shown the superior predictive value of cholesterol efflux functionality of HDL in predicting cardiovascular disease compared to the simple HDL-C levels [41, 42]. One study investigated the relationship between a 9-week exercise program (plus diet) and cholesterol efflux in obese females [43], showing non-significant changes in the cholesterol efflux functional levels, with a significant inverse correlation between weight loss and an increased cholesterol efflux functional level during the intervention period [43]. We conducted a pilot study (with a single arm and small sample design) to see the changes in the cholesterol efflux functional levels among 32 non-obese subjects (mean age 68 years) during a 6-week exercise intervention (Kotani K., Remaley A.T., unpublished data). This population was non-obese. A cholesterol efflux assay was performed based on a previous method [44]: that is, near confluent cells were labelled with $^3$H-cholesterol for 48 hours, washed and effluxed for 18 hours with the indicated lipid acceptors prepared in $\alpha$-MEM containing 1 mg/mL of BSA ($\alpha$-MEM/BSA). The percentage efflux was calculated by subtracting the radioactive count in the blank medium from the radioactive count in the presence of the acceptor and then dividing the result by the sum of the radioactive count in the medium plus the cell fraction. As a result, the overall HDL-C levels increased, and the cholesterol efflux functional levels increased as exercise increased (Table 2). Considering the recent concept of the superiority of the cholesterol efflux functional levels in predicting cardiovascular disease to the simple HDL-C levels, it would be meaningful to note the exercise can improve and/or enhance cholesterol efflux functionality.

**Table 2. Changes of parameters at the pre- and post-intervention**

| Parameters | Pre | Post | P |
|---|---|---|---|
| Body mass index, kg/m$^2$ | 22.9 ± 2.9 | 22.6 ± 3.1 | < 0.01 |
| Glucose, mg/dL | 99 ± 15 | 98 ± 21 | NS |
| Total cholesterol, mg/dL | 215 ± 37 | 230 ± 41 | < 0.01 |
| Triglyceride, mg/dL | 80 (55-112) | 102 (64-124) | NS |
| HDL-C, mg/dL | 64 ± 13 | 68 ± 15 | < 0.01 |
| Cholesterol efflux, % | 21 ± 3 | 26 ± 3 | < 0.01 |

Data are expressed as mean ± standard deviation or median (interquartile range).
NS: no significance, HDL-C: high-density lipoprotein cholesterol.
P: paired t-test (pre and post). Triglyceride values were log-transformed.

## PERSPECTIVES AND CONCLUSION

Circulating HDL-C, represented as the HDL quantity, is a marker of cardioprotection, and exercise generally increases the HDL-C levels. The HDL quality (reflective of HDL functionality) has recently received much attention with respect to cardiovascular health. While less information is available on the association of exercise with HDL quality, we herein presented pilot data regarding the association between exercise and HDL quality based on oxidatively modified HDL and cholesterol efflux functionality of HDL, showing that exercise can favorably affect the HDL quality. This encourages practitioners to take into consideration the view of HDL quality in relation to exercise, besides HDL-C, in terms of the future direction of cardiovascular health. Further factors, such as genetic components, may potentially be considered in order to clarify the association between exercise and the biology of HDL quality [12]. In the near future, with assessments of the HDL quality, the 'quality of exercise' may also be discussed.

## REFERENCES

[1]     Zoungas, S; Curtis, AJ; McNeil, JJ; Tonkin, AM. Treatment of dyslipidemia and cardiovascular outcomes: the journey so far--is this the end for statins? *Clin Pharmacol Ther.*, 2014, 96, 192-205.

[2]     Wenger, NK. Prevention of cardiovascular disease: highlights for the clinician of the 2013 American College of Cardiology/American Heart Association guidelines. *Clin Cardiol.*, 2014, 37, 239-51.

[3]     Rader, DJ; Hovingh, GK. HDL and cardiovascular disease. *Lancet.* 2014, 384, 618-25.

[4]     Subedi, BH; Joshi, PH; Jones, SR; Martin, SS; Blaha, MJ; Michos, ED. Current guidelines for high-density lipoprotein cholesterol in therapy and future directions. *Vasc Health Risk Manag.* 2014, 10, 205-16.

[5]     Rosenson, RS; Brewer, HB; Jr. Davidson, WS; Fayad, ZA; Fuster, V; Goldstein, J; Hellerstein, M; Jiang, XC; Phillips, MC; Rader, DJ; Remaley, AT; Rothblat, GH; Tall, AR; Yvan-Charvet, L. Cholesterol efflux and atheroprotection: advancing the concept of reverse cholesterol transport. *Circulation.* 2012, 125, 1905-19.

[6]     Navab, M; Anantharamaiah, GM; Reddy, ST; Van Lenten, BJ; Ansell, BJ; Fogelman, AM. Mechanisms of disease: proatherogenic HDL--an evolving field. *Nat Clin Pract Endocrinol Metab.* 2006, 2(9), 504-11.

[7]     Vaisar, T; Pennathur, S; Green, PS; Gharib, SA; Hoofnagle, AN; Cheung, MC; Byun, J; Vuletic, S; Kassim, S; Singh, P; Chea, H; Knopp, RH; Brunzell, J; Geary, R; Chait, A; Zhao, XQ; Elkon, K; Marcovina, S; Ridker, P; Oram, JF; Heinecke, JW. Shotgun proteomics implicates protease inhibition and complement activation in the antiinflammatory properties of HDL. *J Clin Invest.*, 2007, 117, 746-56.

[8]     Ferretti, G; Bacchetti, T; Nègre-Salvayre, A; Salvayre, R; Dousset, N; Curatola, G. Structural modifications of HDL and functional consequences. *Atherosclerosis.* 2006, 184(1), 1-7.

[9]     Meilhac, O. High-density lipoproteins in stroke. *Handb Exp Pharmacol.*, 2015, 224, 509-26.

[10]   Tran-Dinh, A; Diallo, D; Delbosc, S; Varela-Perez, LM; Dang, QB; Lapergue, B; Burillo, E; Michel, JB; Levoye, A; Martin-Ventura, JL; Meilhac, O. HDL and endothelial protection. *Br J Pharmacol.*, 2013, 169, 493-511.

[11]   Davidson, WS; Silva, RA; Chantepie, S; Lagor, WR; Chapman, MJ; Kontush, A. Proteomic analysis of defined HDL subpopulations reveals particle-specific protein clusters: relevance to antioxidative function. *Arterioscler Thromb Vasc Biol.*, 2009, 29, 870-6.

[12]   Trejo-Gutierrez, JF; Fletcher, G. Impact of exercise on blood lipids and lipoproteins. *J Clin Lipidol.*, 2007, 1, 175-81.

[13] Blazek, A; Rutsky, J; Osei, K; Maiseyeu, A; Rajagopalan, S. Exercise-mediated changes in high-density lipoprotein: impact on form and function. *Am Heart J.*, 2013, 166(3), 392-400.

[14] Kelley, GA; Kelley, KS; Tran, ZV. Exercise, lipids, and lipoproteins in older adults: a meta-analysis. *Prev Cardiol.*, 2005, 8, 206-14.

[15] Halverstadt, A; Phares, DA; Wilund, KR; Goldberg, AP; Hagberg, JM. Endurance exercise training raises high-density lipoprotein cholesterol and lowers small low-density lipoprotein and very low-density lipoprotein independent of body fat phenotypes in older men and women. *Metabolism.* 2007, 56, 444-50.

[16] Hayashino, Y; Jackson, JL; Fukumori, N; Nakamura, F; Fukuhara, S. Effects of supervised exercise on lipid profiles and blood pressure control in people with type 2 diabetes mellitus: a meta-analysis of randomized controlled trials. *Diabetes Res Clin Pract.*, 2012, 98, 349-60.

[17] Rosenson, RS; Brewer, HB; Jr. Chapman, MJ; Fazio, S; Hussain, MM; Kontush, A; Krauss, RM; Otvos, JD; Remaley, AT; Schaefer, EJ. HDL measures, particle heterogeneity, proposed nomenclature, and relation to atherosclerotic cardiovascular events. *Clin Chem.* 2011, 57, 392-410.

[18] Kelley, GA; Kelley, KS. Aerobic exercise and HDL2-C: a meta-analysis of randomized controlled trials. Atherosclerosis. 2006, 184, 207-15.

[19] Bhalodkar, NC; Blum, S; Rana, T; Bhalodkar, A; Kitchappa, R; Enas, EA. Effect of leisure time exercise on high-density lipoprotein cholesterol, its subclasses, and size in Asian Indians. *Am J Cardiol.* 2005, 96, 98-100.

[20] Muth, ND; Laughlin, GA; von Mühlen, D; Smith, SC; Barrett-Connor, E. High-density lipoprotein subclasses are a potential intermediary between alcohol intake and reduced risk of cardiovascular disease: the Rancho Bernardo Study. *Br J Nutr.* 2010, 104, 1034-42.

[21] Campbell, SC; Moffatt, RJ; Kushnick, MR. Continuous and intermittent walking alters HDL(2)-C and LCATa. *Atherosclerosis.* 2011, 218, 524-9.

[22] Superko HR; Pendyala, L; Williams, PT; Momary, KM; King, SB; 3rd Garrett, BC. High-density lipoprotein subclasses and their relationship to cardiovascular disease. *J Clin Lipidol.*, 2012, 6, 496-523.

[23] Calabresi, L; Gomaraschi, M; Franceschini, G. High-density lipoprotein quantity or quality for cardiovascular prevention? *Curr Pharm Des.*, 2010, 16, 1494-503.

[24] Kingwell, BA; Chapman, MJ; Kontush, A; Miller, NE. HDL-targeted therapies: progress, failures and future. *Nat Rev Drug Discov.*, 2014, 13, 445-64.

[25] Tsompanidi, EM; Brinkmeier, MS; Fotiadou, EH; Giakoumi, SM; Kypreos, KE. HDL biogenesis and functions: role of HDL quality and quantity in atherosclerosis. *Atherosclerosis.*, 2010, 208,3-9.

[26] Karavia, EA; Zvintzou, E; Petropoulou, PI; Xepapadaki, E; Constantinou, C; Kypreos, KE. HDL quality and functionality: what can proteins and genes predict? *Expert Rev Cardiovasc Ther.*, 2014, 12, 521-32.

[27] Katsiki, N; Athyros, VG; Karagiannis, A; Mikhailidis, DP. High-density lipoprotein, vascular risk, cancer and infection: a case of quantity and quality? *Curr Med Chem.*, 2014, 21, 2917-26.

[28] Remaley, AT; Warnick, GR. High-density lipoprotein: what is the best way to measure its antiatherogenic potential? *Expert Opin Med Diagn.*, 2008, 2, 773-88.

[29] Pankhurst, G; Wang, XL; Wilcken, DE; Baernthaler, G; Panzenböck, U; Raftery, M; Stocker, R. Characterization of specifically oxidized apolipoproteins in mildly oxidized high density lipoprotein. *J Lipid Res.*, 2003, 44, 349-55.

[30] Zheng, L; Nukuna, B; Brennan, ML; Sun, M; Goormastic, M; Settle, M; Schmitt, D; Fu, X; Thomson, L; Fox, PL; Ischiropoulos, H; Smith, JD; Kinter, M; Hazen, SL. Apolipoprotein A-I is a selective target for myeloperoxidase-catalyzed oxidation and functional impairment in subjects with cardiovascular disease. *J Clin Invest.*, 2004, 114, 529-41.

[31] Shao, B; Oda, MN; Vaisar, T; Oram, JF; Heinecke, JW. Pathways for oxidation of high-density lipoprotein in human cardiovascular disease. *Curr Opin Mol Ther.*, 2006, 8, 198-205.

[32] von Eckardstein, A; Walter, M; Holz, H; Benninghoven, A; Assmann, G. Site-specific methionine sulfoxide formation is the structural basis of chromatographic heterogeneity of apolipoproteins A-I, C-II, and C-III. *J Lipid Res.*, 1991, 32, 1465-76.

[33] Bowry, VW; Stanley, KK; Stocker, R. High density lipoprotein is the major carrier of lipid hydroperoxides in human blood plasma from fasting donors. *Proc Natl Acad Sci U S A.* 1992, 89, 10316-20.

[34] Francis, GA. High density lipoprotein oxidation: in vitro susceptibility and potential in vivo consequences. *Biochim Biophys Acta.* 2000, 1483, 217-35.

[35] Nakano, T; Nagata, A. Oxidative susceptibility of apolipoprotein AI in serum. *Clin Chim Acta.*, 2005, 362, 119-24.

[36] Navab, M; Reddy, ST; Van Lenten, BJ; Fogelman, AM. HDL and cardiovascular disease: atherogenic and atheroprotective mechanisms. *Nat Rev Cardiol.*, 2011, 8, 222-32.

[37] Ueda, M; Hayase, Y; Mashiba, S. Establishment and evaluation of 2 monoclonal antibodies against oxidized apolipoprotein A-I (apoA-I) and its application to determine blood oxidized apoA-I levels. *Clin Chim Acta.*, 2007, 378, 105-11.

[38] Kotani, K; Sakane, N; Ueda, M; Mashiba, S; Hayase, Y; Tsuzaki, K; Yamada, T; Remaley, AT. Oxidized high-density lipoprotein is associated with increased plasma glucose in non-diabetic dyslipidemic subjects. *Clin Chim Acta.*, 2012, 414, 125-9.

[39] Ueda, M; Hayase, Y; Mashiba, S. Establishment and evaluation of 2 monoclonal antibodies against oxidized apolipoprotein A-I (apoA-I) and its application to determine blood oxidized apoA-I levels. *Clin Chim Acta.* 2007, 378, 105-11.

[40] Huang, Y; DiDonato, JA; Levison, BS; Schmitt, D; Li, L; Wu, Y; Buffa, J; Kim, T; Gerstenecker, GS; Gu, X; Kadiyala, CS; Wang, Z; Culley, MK; Hazen, JE; Didonato, AJ; Fu, X; Berisha, SZ; Peng, D; Nguyen, TT; Liang, S; Chuang, CC; Cho, L; Plow, EF; Fox, PL; Gogonea, V; Tang, WH; Parks, JS; Fisher, EA; Smith, JD; Hazen, SL. An abundant dysfunctional apolipoprotein A1 in human atheroma. *Nat Med.*, 2014, 20, 193-203.

[41] Khera, AV; Cuchel, M; de la Llera-Moya, M; Rodrigues, A; Burke, MF; Jafri, K; French, BC; Phillips, JA; Mucksavage, ML; Wilensky, RL; Mohler, ER; Rothblat, GH; Rader, DJ. Cholesterol efflux capacity, high-density lipoprotein function, and atherosclerosis. *N Engl J Med.*, 2011, 364, 127-35.

[42] Rohatgi, A; Khera, A; Berry, JD; Givens, EG; Ayers, CR; Wedin, KE; Neeland, IJ; Yuhanna, IS; Rader, DR; de Lemos, JA; Shaul, PW. HDL cholesterol efflux capacity and incident cardiovascular events. *N Engl J Med.* 2014, 371, 2383-93.

[43] Králová Lesná, I; Suchánek, P; Kovár, J; Poledne, R. Life style change and reverse cholesterol transport in obese women. *Physiol Res.*, 2009, 58, S33-8.

[44] Remaley, AT; Schumacher, UK; Stonik, JA; Farsi, BD; Nazih, H; Brewer, HB. Jr. Decreased reverse cholesterol transport from Tangier

disease fibroblasts. Acceptor specificity and effect of brefeldin on lipid efflux. *Arterioscler Thromb Vasc Biol.*, 1997, 17, 1813-21.

In: Blood Lipids and Lipoproteins
Editor: Melissa R. Ruiz
ISBN: 978-1-63482-591-7
© 2015 Nova Science Publishers, Inc.

*Chapter 2*

# NATURALLY OCCURRING MULTIPLE-MODIFIED LOW DENSITY LIPOPROTEIN

*Alexander N. Orekhov[1,2,3]\*, Ekaterina A. Ivanova[4]*
*and Yuri V. Bobryshev[1,5]#*

[1]Institute of General Pathology and Pathophysiology, Moscow, Russia
[2]Institute for Atherosclerosis Research, Moscow, Russia
[3]Department of Biophysics, Biological Faculty,
Moscow State University, Moscow, Russia
[4]Department of Pediatric Nephrology & Growth and Regeneration,
Katholieke Universiteit Leuven and University Hospitals Leuven,
Leuven, Belgium
[5]Faculty of Medicine, School of Medical Sciences,
University of New South Wales, Sydney NSW, Australia

## ABSTRACT

Extra- and intracellular deposition of lipids, predominantly of cholesteryl esters, in arterial intima is one of the earliest manifestations of atherosclerosis. Formation of lipid-laden foam cells is recognized as a trigger in the pathogenesis of atherosclerosis. Low-density lipoprotein (LDL) circulating in human blood is the source of lipids accumulated in

\* Corresponding authors: Alexander N. Orekhov (a.h.opexob@gmail.com).
# Dr Yuri V Bobryshev, Faculty of Medicine, University of New South Wales, Kensington NSW
2052, Australia, Phone/Fax: (6 12) 9385 1217, E-mail: y.bobryshev@unsw.edu.au.

arterial cells. However, numerous attempts to induce intracellular accumulation of cholesteryl esters by native LDL failed. On the other hand, LDL chemically modified in vitro (acetylated, malondialdehyde-treated, oxidized with ions of transient metals, etc.) induced lipid deposition in cells, i.e., is atherogenic. At the same time, the attempts to find LDL in circulation, similar to in vitro modified LDL, were unsuccessful. Thus, a paradoxical situation existed: 1) LDL is the source of lipids accumulated in the vascular wall; 2) native LDL does not induce accumulation of intracellular fat (foam cell formation); 3) in vitro modified forms of LDL are atherogenic, but nobody found them in circulation. In response to this challenge, intensive search for modified forms of circulating LDL have been started, and such forms were soon found. Among modified forms of LDL detected in blood, such forms as oxidized, small dense, and electronegative have been described. We discovered modified desialylated LDL in human blood plasma. This LDL was capable of inducing lipid accumulation in cultured cells. We isolated atherogenic LDL and characterized it. It was shown that atherogenic LDL possesses numerous alterations of carbohydrate, protein, and lipid moieties, and can be termed multiple-modified LDL. Multiple modification of LDL occurs in human blood plasma and represents a cascade of successive changes in the lipoprotein particle: desialylation, loss of lipids, reduction in the particle size, increase of surface electronegative charge, etc. In addition to intracellular lipid accumulation, stimulatory effects of naturally occurring multiple-modified LDL on other processes involved in the development of atherosclerotic lesions, namely cell proliferation and fibrosis, were shown.

**Keywords:** Atherogenicity, atherosclerosis, autoantibodies, desialylation, electronegative, intracellular lipid accumulation, low-density lipoprotein, oxidation, small dense, trans-sialidase

# INTRODUCTION

Extra- and intracellular deposition of lipids, predominantly of cholesteryl esters, in arterial intima is one of the earliest manifestations of atherosclerosis [1-4]. Formation of lipid-laden foam cells is recognized as the initiating moment in the pathogenesis of atherosclerosis [5, 6]. By the end of 1970s, it was found that low-density lipoprotein (LDL) circulating in human blood is the source of lipids accumulated in vascular cells [7, 8]. However, numerous attempts to induce intracellular accumulation of cholesteryl esters by native LDL failed [9, 10]. On the other hand, LDL chemically modified in vitro (acetylated, malondialdehyde-treated, oxidized with ions of transient metals, etc.) induced lipid deposition in cells, i.e., is atherogenic [11-14]. At the same time, the attempts to find LDL in circulation similar to in vitro modified LDL were unsuccessful. Thus, a paradoxical situation existed in the early 1980s: 1) LDL is the source of lipids accumulated in the vascular wall; 2) native LDL does not induce accumulation of intracellular fat (foam cell formation); 3) in vitro modified forms of LDL are atherogenic, but nobody found them in circulation. In response to this challenge, intensive search for modified forms of circulating LDL started and such forms were soon found.

# 1. OXIDIXED LDL

Currently, PubMed lists 7899 articles indexed under "oxidized LDL" and 3682 under "oxidized LDL and atherosclerosis." Hundreds of reviews were written on this topic so there is no need to dwell on the oxidative modification of LDL in detail. The idea of the crucial role of oxidized LDL in atherogenesis was promoted primarily by Steinberg and co-workers, who established that LDL incubated with cultured cells undergoes free radical-catalyzed oxidative modification that generates lipid peroxides and extensive structural changes in the LDL molecule [13].

Others have reported that incubation with oxidized LDL resulted in a significant increase in cholesterol ester, while incubation with native LDL did not result in cholesterol accumulation in cultured mouse peritoneal macrophages [15].

At present, it is generally accepted that oxidized LDL causes accumulation of lipids in the arterial wall and triggering atherogenesis [16-18]. However, oxidized LDL itself had not (and still has not) been demonstrated in blood. In this regard, it has been suggested that oxidation of LDL occurs not in the blood but in the arterial wall where oxidized LDL manifests its atherogenic action. The ideas on the key role of oxidized LDL in atherogenesis are based on the following arguments.

1.   In the blood, autoantibodies against malondialdehyde-LDL (MDA-LDL) have been found [19].
2.   Antibodies against artificially oxidized LDL recognize materials in the atherosclerotic lesions where LDL is also co-localized with oxidation products [20].

3.  At least a portion of the LDL isolated from atherosclerotic lesions is similar, if not identical, to oxidatively modified LDL [21].

It should be emphasized that circulating antibodies have affinity not to oxidized LDL but to MDA-LDL, a model of oxidized LDL; however, these antibodies have the higher affinity to desialylated LDL [22]. Thus, anti-LDL autoantibodies that primarily react with desialylated LDL show cross-reactivity with MDA-LDL. Obviously, autoantibodies are produced in response to the appearance of desialylated LDL but not oxidized LDL. This and other facts lead to doubt that the oxidative modification of LDL is the only oт vivo atherogenic modification responsible for the onset and progression of atherosclerotic lesions. Especially since in the blood other existing atherogenic modifications of LDL have been found that undeservedly are paid much less attention.

## 2. SMALL DENSE LDL

LDL isolated by various techniques is defined as lipoprotein fraction with density range from 1.019 to 1.063 g/l. Ultracentrifugation and gradient gel electrophoresis are widely used for analysis of LDL subfractions. Using these methods, LDL particles are classified into several subclasses, including large, intermediate, small, very small LDL [23, 24].

The first method that has been used for separation of different LDL subfractions was analytical Ultracentrifugation [24-27]. In this method, LDL particles are separated based on their flotation rate. Small and very small LDLs are usually referred to as small dense LDL (sdLDL).

Non-denaturating gradient gel electrophoresis separates LDL subfractions by their electrophoretic mobility, which is determined by the size and shape of the particle [28]. Gradient gel electrophoresis separation reveals 4 subclasses: large, intermediate, small, and very small LDL [29]. There is a strong correlation between LDL size and density of particles revealed by ultracentrifugation and gradient gel electrophoresis, however, these parameters are not identical. In some studies tube gel electrophoresis is used for LDL subfraction analysis to obtain quantitative results [30].

Nuclear magnetic resonance is used for analysis of LDL subfractions in blood plasma, however, the results obtained by nuclear magnetic resonance differ significantly from data obtained by gradient gel electrophoresis, and cannot be directly compared [31, 32]. Other methods can be employed for

LDL subfraction analysis including high-performance liquid chromatography [33], dynamic light scattering [34, 35], ion mobility analysis [36, 37] and homogenous assay analysis [38]. The standardization is a problem for clinical and diagnostic estimation of LDL subfractions because different methods of LDL subfraction analysis demonstrate different results. Thus, we need more studies to create a standard analytical technique.

The exact origins of LDL subfractions remain not fully clear. According to Berneis and Krauss, two types of precursors are secreted from liver, namely triglyceride-rich and triglyceride-poor apolipoprotein B (apoB) [39]. TG-poor lipoprotein is a precursor for large LDL subfractions, whereas TG-rich lipoprotein is a precursor for sdLDL subfraction. This theory explains the metabolic pathway for sdLDL from liver-secreted precursors, and is supported by clinical investigations [39, 40].

Genetic factors affecting sdLDL production have been studied in genome-wide association studies (GWAS). It was found that sdLDL metabolism is dependent on genetic factors that might be considered for the development of novel therapeutic strategies [37, 41].

The lifetime of sdLDL is longer than that of large LDL that is removed from the bloodstream via the LDL receptor pathway [42, 43]. Intracellular cholesterol accumulation with foam cell formation is the key process that leads to the development and growth of atherosclerotic lesions in the arterial wall. LDL is the main source accumulated. Many researchers reported that native LDL does not cause lipid accumulation in cultured cells, whereas modified LDL is highly atherogenic [44, 45]. Oxidation is one of the atherogenic modifications of LDL that have been proposed [44]. It was reported that sdLDL contains less vitamins antioxidants and is therefore more susceptible to oxidation than larger forms of LDL [46].

In vivo atherogenic modification of LDL is desialylation, which is performed in blood by trans-sialidase [47]. Trans-sialidase transfers terminal sialic acid from the LDL particle to various plasma acceptors. Incubation of native LDL with blood plasma of atherosclerotic patients leads to a gradual desialylation of LDL particles [48]. sdLDL has a decreased sialic acid content in comparison to larger LDL particles [49]. Desialylation apparently increases sdLDL binding to arterial proteoglycans that leads to retention of desialylated sdLDL in the subendothelial space where it can contribute to the atherosclerosis plaque development [50].

# 3. ELECTRONEGATIVE LDL

Another form of circulating modified LDL is electronegative LDL (LDL(-)), which can be revealed by agarose gel electrophoresis, isotachophoresis or ion exchange chromatography [51, 52]. Atherogenic LDL(-) fraction was first isolated by Avogaro and co-authors [52] using ion-exchange chromatography.

Five subfractions of LDL(-) with various degrees of electronegativity were described [53, 54]. Plasma levels of the most electronegative subfraction were associated with cardiovascular risks, including smoking, hyper-cholesterolemia, type 2 diabetes, and myocardial infarction [55-57].

In addition to anion exchange chromatography the capillary isotachophoresis is used for LDL(-) isolation and analysis [58] This analytical method discriminates LDL(-) as fast migrating fraction of LDL. Capillary isotachophoresis was used to analyze the LDL(-) in heparin precipitate [59]. Another method based on monoclonal antibodies distinguishes LDL(-) by the specific epitopes [60-62]. This approach allows to develop LDL(-) ELISA useful for clinical practice [63].

LDL(-) particles possess the tendency to aggregate [52]. Aggregation of LDL(-) was investigated in details [64]. As is known, association of LDL particles leads to their atherogenicity, i.e., ability of LDL to cause accumulation of intracellular cholesterol [65-68]. The misfolded lipoprotein is the decisive factor responsible for LDL(-) aggregation [69]. Secondary structure of apoB is violated in LDL(-) [70]. Misfolded conformation of apoB in LDL(-) is characterized by abnormal exposure of tryptophan residues to the aqueous environment [64, 70]. Using the 2D-NMR analysis it was demonstrated that a number of lysine residues in LDL(-) have altered ionization status [71]. Changes in lipid composition of LDL(-) particles alter their surface properties, thus may also contribute to its aggregation ability [72-74].

Misfolded conformation of protein structure of the LDL(-) particles is a cause of decreased affinity to specific LDL receptors and prolonged circulation times of LDL(-) [75, 76]. However, the most electronegative fraction of LDL(-) binds to the lectin-like oxidized LDL receptor 1 (LOX-1) [55, 77]. The most electronegative subfraction of LDL(-) induces atherogenic cellular manifestations, namely it elevates production of reactive oxygen species and increases the C-reactive protein levels in cultured endothelial cells via LOX-1 signalling [78].

Thus, LDL(-) can activate inflammatory and immune responses that contribute to the atherogenesis. Due to binding to proteoglycans, LDL(-) retention takes place in the subendothelial intima. Subendothelial cells take up LDL(-) through scavenger receptors turning into foam cells. Anti-LDL(-) autoantibodies may also participate in atherogenesis [79]. LDL(-) possesses cytotoxic effect on endothelial cells, causing apoptosis and production of inflammatory molecules such as IL-8, MCP-1 and VCAM-1 [79-81]. Therefore, the accumulated data revealed pro-atherogenic and pro-inflammatory properties of LDL (-).

## 4. DISCOVERY AND ISOLATION OF CIRCULATING DESIALYLATED MODIFIED LDL

If someone wants to find modified LDL in the blood then the most reasonable way to search for them is in the blood of patients with atherosclerosis. As a first step, we isolated LDL from blood of healthy subjects and patients with angiographically assessed atherosclerosis. Our objective was to find that LDL is capable of causing lipid accumulation in the arterial cells. The ability of LDL to induce intracellular lipid accumulation was tested in a primary culture of human aortic intima smooth muscle α-actin-positive cells (typical smooth muscle cells and pericyte-like cells), those cells that accumulate fat in atherosclerotic lesions in situ [82]. The method of isolation and cultivation of these cells was developed in our laboratory [82]. On the 7th day in culture, smooth muscle cells α-actin-positive cells (SMA(+) cells) isolated from uninvolved human aortic intima were incubated for 24 h in Medium 199 containing 10% lipoprotein-deficient serum of healthy donor and LDL samples at a concentration of 5-500 μg apoprotein B (apo B)/ml. In most cases the LDL samples obtained from healthy individuals induced no intracellular accumulation of phospholipids and neutral lipids [83, 84], while most samples of LDL isolated from the plasma of patients with coronary atherosclerosis, increased intracellular contents of free cholesterol and triglycerides by 1.5-fold, and that of cholesteryl esters by 1.5- to 5-fold. Further increase in LDL concentration did not lead to additional increase of intracellular lipid level. Therefore, we have shown that the majority of LDL samples obtained from the blood of atherosclerotic patients but not healthy individuals causes accumulation of lipids in human vascular cells. This property of LDL was termed atherogenicity [85]. What is the reason for LDL atherogenicity? We have attempted to find the differences between atherogenic LDL circulating in the blood of patients and nonatherogenic LDL of healthy subjects. We were lucky since when comparing atherogenic and nonatherogenic LDL, we found a significant difference in the content of sialic acid. The total content of sialic acid (N-acetylneuraminic acid) in LDL from coronary atherosclerosis patients was 2- to 3-fold lower than that in LDL from healthy subjects. LDL with low sialic acid content was primarily termed desialylated LDL [86, 87]. Sialic acid is a terminal monosaccharide of asparagine-bound biantennary carbohydrate chains of LDL glycoconjugate moiety. After removal of sialic acid, galactose becomes the terminal saccharide. We took advantage of this fact to isolate the subfraction of desialylated LDL from total LDL preparation using Ricinus communis agglutinin (RCA120), which possesses a high affinity to the terminal galactose [88]. Total LDL preparation was applied on column with RCA120 immobilized on CNBr-activated agarose. Normally sialylated LDL passed through the column not binding to the sorbent. Desialylated LDL was bound to lectin and then eluted with 5-50 mM galactose. This approach allowed us to isolate subfactions of both sialylated and desialylated LDL from total LDL preparation derived from the blood of patients. Besides, we found that desialylated particles of LDL represent only some portion of all LDL particles circulating in the blood of patients. Using lectin affinity chromatography and lectin-sorbent assay, we have shown that the proportion of desialylated LDL in patients' blood varies from 20% to 60% of the total LDL level, while in healthy subjects it varies from 5 to 15% [89]. The sialic acid content of desialylated LDL fraction isolated by lectin chromatography was 2- to 3-fold lower than that of sialylated LDL [88]. As the further step, we tried to assess the atherogenic potential of sialylated and desialylated LDL subfractions. Incubation of cultured SMA(+) cells obtained from human aortic intima with sialylated LDL subfraction had no effect on intracellular content of phospholipids and neutral lipids [88, 90]. Desialylated LDL subfraction induced 1.5- to 2-fold increase in the intracellular content of unesterified cholesterol and triglycerides and a 2- to 7-fold increase in that of cholesteryl esters. Hence, only desialylated subfraction of human LDL is atherogenic. Sialylated LDL possesses no atherogenic activity and can be defined as native unmodified LDL. Thus, we have identified and isolated a subfraction of naturally occurring modified (desialylated) LDL capable of inducing lipid accumulation in human arterial subendothelial cells.

# 5. METABOLISM OF NATURALLY OCCURRING MODIFIED (DESIALYLATED) LDL IN SMA(+) CELLS OF HUMAN AORTIC INTIMA

Two approaches were used to elucidate the mechanisms of intracellular lipid accumulation caused by desialylated LDL: 1) evaluation of binding, uptake and degradation of [$^{125}$I]LDL in SMA(+) cells of normal and atherosclerotic human aortic intima and 2) determination of the rate of hydrolysis and esterification of radiolabeled lipids incorporated in LDL particles.

## 5.1. Binding And Uptake of [$^{125}$I ] LDL by Human Aortic Intimal SMA(+) Cells

First, we studied the uptake of native LDL and desialylated LDL by cells cultured from uninvolved and atherosclerotic (fatty streaks and atherosclerotic plaques) human aortic intima. The uptake of iodinated native LDL by cells cultured from normal and atherosclerotic intima was similar (Table 1). The uptake of desialylated LDL by cells cultured from uninvolved intima was 2.5-fold higher than that of native LDL. The uptake of desialylated LDL by cells cultured from fatty streaks and atherosclerotic plaques was 6-fold higher as compared to native LDL. Thus, the uptake of desialylated LDL was much higher than the uptake of native LDL, especially by cells cultured from atherosclerotic lesions.

We have suggested that the enhanced uptake of desialylated LDL is associated with elevated binding of this lipoprotein to the cell surface. In fact, the binding of desialylated LDL to cells cultured from uninvolved and atherosclerotic intima was 2- to 3-fold higher as compared to native LDL (Table 1).

Scatchard analysis of LDL binding showed that the binding of desialylated LDL is unsaturated, which indicates the presence of additional low affinity binding sites [92]. We have found that cellular binding of iodinated desialylated LDL was only partially inhibited by a 20-fold excess of unlabeled native LDL and by antibodies against the native LDL receptor. These data confirm the presence of additional binding sites for modified LDL.

Availability of terminal galactose in desialylated LDL suggests that modified LDL can bind to the asialoglycoprotein receptor. It was established that 50 mM galactose and neuraminidase-treated fetuin inhibit the binding of desialylated LDL to smooth muscle cells cultured from human aortic intima. These agents did not inhibit the binding of native LDL to the cells. These data show the ability of desialylated LDL to interact with asialoglycoprotein receptor of vascular SMA(+) cells.

To test the ability of desialylated LDL to bind to scavenger-receptor of aortic intimal SMA(+) cells, [$^{125}$I]LDL was incubated with a 20-fold excess of unlabeled acetylated LDL. Acetylated LDL inhibited the binding of desialylated LDL by 35% and had no effect on the binding of native LDL. On the other hand, a 20-fold excess of unlabeled desialylated LDL inhibited the binding of acetylated [$^{125}$I]LDL by 30%. So, desialylated LDL can compete with acetylated LDL for the scavenger-receptor. Moreover, desialylated LDL interaction with dimer and trimer of scavenger-receptor of human THP-1 macrophages was demonstrated by immunoblotting [90]. Thus, desialylated LDL interacts with scavenger-receptor of smooth muscle cells and macrophages.

### Table 1. Binding, Uptake and Degradation of Native LDL and Desialylated LDL by SMA(+) Cells Cultured from Undiseased and Atherosclerotic Intima of Human Aorta

| LDL | LDL Binding | LDL Uptake | LDL Degradation |
|---|---|---|---|
| | ng/mg of cell protein | | |
| | Undiseased intima | | |
| Native LDL | 95±4 | 1148±84 | 6248±345 |
| Desialylated LDL | 178±6* | 2658±187* | 3905±307* |
| | Fatty Streaks | | |
| Native LDL | 98±6 | 1215±106 | 4148±189 |
| Desialylated LDL | 315±21* | 6178±187* | 1764±168* |
| | Atherosclerotic plaques | | |
| Native LDL | 97±8 | 1108±109 | 4207±223 |
| Desialylated LDL | 307±20* | 6095±194* | 1605±93* |

*, Significantly different from native LOL, p<O.05.

To evaluate the binding of LDL to cellular proteoglycans, human intimal SMA(+) cells were treated with enzymes hydrolyzing their polysaccharide chains (hyaluronidase, heparinase-heparitinase mixture and chondroitinase). Treatment of the cells with hyaluronidase had no effect on the binding of native LDL to normal and atherosclerotic cells. On the other hand, the binding of desialylated LDL to hyaluronidase-treated cells was lower than that of intact cells. Treatment of cells with heparinase-heparitinase or chondroitinase ABC reduced cellular binding of desialylated LDL but not of native LDL. The addition of lipoprotein lipase forming complexes with both proteoglycans and LDL to the incubation medium did not change lipoprotein binding. These findings indicate that in contrast to native LDL, desialylated LDL binds to cell surface proteoglycans and this binding is not mediated by lipoprotein lipase.

Thus, the binding to the scavenger-receptor, asialoglycoprotein-receptor and proteoglycans may account for enhanced cellular binding and uptake of desialylated LDL.


## 5.2. Intracellular Degradation of Apolipoprotein B

Degradation rate of desialylated LDL in normal cells was 1.5-fold lower than that of native LDL (Table 1). The difference in the degradation rate of native LDL and desialylated LDL was more pronounced in cells cultured from tatty streaks and atherosclerotic plaques. To elucidate the causes of low rate of desialylated LDL degradation we used homogenates of SMA(+) cells isolated from normal and atherosclerotic human aortic intima to study proteolysis at different pH levels as well as the effects of agents that are poorly internalized by intact cells, and to determine the activities of proteolytic enzymes [92]. In cell homogenates the degradation rate of [$^{125}$I]apoB of desialylated LDL was slower than that of native LDL. Two optima at pH 4.5 and pH 5.5 have been revealed for apoB hydrolysis. The difference in the rate of degradation of native LDL and desialylated LDL has been demonstrated. The difference in the rate of degradation of native and desialylated LDL at these pH optima has been demonstrated. This suggests that different proteinases are involved in the degradation of apolipoproteins of native LDL and desialylated LDL.

To identify proteinases involved in apoB degradation, we used the inhibitors of major lysosomal proteinases. Inhibitors of thiol (iodoacetamide), serine (soybean trypsin inhibitor, phenylmethanesulfonyl fluoride) and metalloproteinases (1,10-phenantroline), as well as of cathepsins A and B (leupeptine and antipaine), inhibited hydrolysis of apoB of native LDL and desialylated LDL by 20-80%. It was found that cathepsin A is involved in degradation of apolipoprotein B of desialylated LDL but not native LDL. Pepstain A, an inhibitor of the cathepsin D (carboxyl proteinase), completely inhibited degradation of apoB of both native LDL and desialylated LDL. Kinetic studies showed that primary hydrolysis of apoB by cathepsin D is a necessary stage for degradation of the apolipoprotein by other proteinases. Determination of proteolytic hydrolysis in homogenates of cells isolated from fatty streaks and atherosclerotic plaques showed that the activities of cathepsins D, B and A were 1.5- to 2-fold lower than in homogenates of normal cells [92]. Thus, both in normal and atherosclerotic cells, the rate of degradation of internalized desialylated LDL is lower than that of native LDL. The enhanced uptake and low rate of intracellular degradation lead to accumulation of desialylated LDL in SMA(+) cells of human aortic intima. Accumulation of lipoprotein particles with intact or partially degraded apolipoprotein results in appearance of a considerable intracellular pool of lipids (primarily, free and esterified cholesterol) that may serve as the source for the lipid droplet formation.

## 5.3. Intracellular Metabolism of Lipoprotein Lipids

To study the catabolism of cholesterol esters, [$^3$H]cholesteryl linoleate was incorporated in native LDL and desialylated LDL particles [92]. After a 24-h incubation with labeled desialylated LDL, the intracellular content of [$^3$H]cholesteryl linoleate was 2-fold higher than after incubation with native LDL. For the cells cultured from atherosclerotic lesions, the 3- to 5-fold difference was observed. The proportion of hydrolyzed [$^3$H]cholesteryl linoleate of native LDL was 80% both in normal and atherosclerotic cells. The proportion of hydrolyzed [$^3$H]cholesteryl linoleate of desialylated LDL was 40% for normal cells and 25% for atherosclerotic cells. So, the rate of hydrolysis of desialylated LDL cholesteryl esters in SMA(+) cells from normal and atherosclerotic human aortic intima is considerably lower compared with that of native LDL.

In order to determine the rate of cholesterol esterification SMA(+) cells were incubated in the presence of [$^{14}$C]oleic acid [92] . The rate of cholesterol esterification in normal and atherosclerotic cells incubated with desialylated LDL was 2- to 3-fold higher than in the case of native LDL.

Thus, desialylated LDL stimulates intracellular esterification of free cholesterol. Obtained data can explain mechanisms of accumulation of cholesteryl esters in human aortic intimal SMA(+) cells caused by desialylated LDL.

## 6. EFFECT OF NATURALLY OCCURRING MODIFIED (DESIALYLATED) LDL ON PROLIFERATIVE ACTIVITY AND SYNTHESIS OF THE EXTRACELLULAR MATRIX COMPONENTS

In addition to intracellular lipid accumulation, increased proliferative activity and enhanced synthesis of the extracellular matrix components by subendothelial cells are generally recognized as major manifestations of atherosclerosis at the cellular level [93]. We decided to examine the effect of circulating desialylated LDL on proliferative activity and synthesis of total protein, collagen and glycosaminoglycans by SMA(+) cells cultured from uninvolved human aortic intima.

A 24-h incubation of SMA(+) cells in Medium 199 containing 10% fetal calf serum and 100 µg/ml native LDL had no effect on [$^3$H]thymidine incorporation [93]. By contrast, addition of the same concentration of desialylated LDL subfraction leads to a 1.5- to 2-fold increase of proliferative activity.

The rate of synthesis of proteins secreted by cultured cells was evaluated by incorporation of [$^{14}$C] leucine in the acid-insoluble fraction of culture medium.

Native LDL had no effect on the synthesis of secreted proteins. Desialylated LDL stimulated protein syntheses by 1.5- to 2-fold [93]. Moreover, desialylated LDL induced a 2-fold increase of collagen production, as estimated by incorporation of [$^{14}$C]proline in the collagenase-released fraction of culture medium [93]. It was also demonstrated that desialylated LDL, but not native LDL, stimulates the incorporation of [$^{14}$C] glucosamine in the total glycosaminoglycan fraction of human SMA(+) cells [93]. So, in contrast to native LDL, desialylated LDL enhanced synthesis of connective tissue matrix components.

The investigation of the mechanisms underlying the stimulatory effect of desialylated LDL on proliferative and synthetic activities of cultured SMA(+) cells showed that: 1) preincubation of cells with desialylated LDL inducing accumulation of intracellular lipids is sufficient to stimulate the incorporation of radiolabeled thymidine and precursors of the extracellular matrix components; 2) insoluble complexes of native LDL with naturally occurring (collagen, elastin, fibronectin) and artificial (latex particles, dextran sulfate) compounds induce intracellular lipid accumulation and stimulate proliferative and synthetic activities of cultured cells; 3) increase in proliferative and synthetic activities correlates with the amount of accumulated intracellular cholesterol [93, 94].

Thus, intracellular lipid accumulation induced by desialylated LDL causes the enhanced proliferative activity and synthesis of the connective tissue matrix components.

Therefore, desialylated LDL can induce all known atherosclerotic manifestations at the cellular level.

# 7. CHARACTERIZATION OF NATURALLY OCCURRLNG MODIFIED (DESIALYLATED) LDL

We have studied different properties of desialylated LDL subfraction in comparison with native LDL.

## 7.1. Chemical Composition

Human apolipoprotein B glycoconjugates contain N-acetyl-glucosamine, galactose, mannose and sialic acid in a molar ratio 2: 1:2.5: 1 [95]. The content of N-acetyl-glucosamine, galactose and mannose in protein-bound glycoconjugates of native LDL and desialylated LDL from healthy subjects was similar. The levels of N-acetyl-glucosamine in apoB of native and desialylated LDL subfractions isolated from coronary atherosclerosis patients were similar and did not differ from those in the lipoprotein subfractions of healthy individuals. The galactose and mannose content in desialylated LDL from patients was significantly lower than in native LDL [96].

The sialic acid level in desialylated LDL from healthy subjects was lower by 15-30% than that in native LDL [96]. The content of sialic acid in apoB of native LDL from atherosclerotic patients was similar to that of native LDL from healthy subjects. The amount of sialic acid in protein-bound glycoconjugates of desialylated LDL was 2- to 3-fold lower than that of native LDL [96].

The carbohydrate composition lipid moiety of LDL differs from the composition of protein-bound saccharide chains by the absence of mannose and the presence of N-acetylgalactosamine and glucose. Some LDL samples contained trace amounts of fucose. The content of lipid-bound N-acetyl-glucosamine was 5- to 9-fold lower, while the content of galactose was

1.5- to 2-fold higher than in apoB glycoconjugales. The amount of lipid-bound sialic acid was 3- to 5-fold lower than in protein-bound carbohydrate chains [96]. The content of all neutral monosaccharides (N-acetyl-galactosamine, N-acetyl-glucosamine, galactose and glucose) in lipid-bound glycoconjugates of desialylated LDL isolated from healthy subjects was 1.5-fold lower than in native LDL [95]. The content of lipid-bound N-acetyl-galactosamine and N-acetyl-glucosamine in desialylated LDL was lower than in native LDL, while the levels of galactose and glucose were similar. The amount of both neutral saccharides and sialic acid in the lipid moiety of desialylated LDL was 1.5- to 3-fold lower than that in native LDL.

The content of major neutral lipids in native and desialylated LDL was determined [92, 96]. The levels of free and esterified cholesterol and triglycerides in desialylated subfraction of LDL were lower by 30-40% as compared to native LDL. The levels of monoglycerides and free fatty acids in desialylated LDL were 1.5-fold higher than in native LDL. The content of free and esterified cholesterol as well as triglycerides in desialylated LDL was 1.5- to 2-fold lower than native LDL. The levels of mono- and diglycerides as well as free fatty acids in desialylated LDL were 3- to 4-fold lower than in native LDL.

In desialylated LDL the content of phosphatidylcholine, phosphate-dylethanolamine and sphingomyelin was lower, while the content of lysophosphatidylcholine was higher than in native LDL [92, 96]. The levels of main phospholipid classes in native LDL did not differ from those in native LDL. However, the content of phosphatidylcholine, phosphatidylethanolamine and sphingomyelin in desialylated LDL was 1.5- to 2-fold higher than in native LDL. The content of phosphatidylinositol and phosphatidylserine in desialylated and native LDL fractions was similar.

The obtained data indicate that desialylated LDL differs considerably from native LDL with respect to carbohydrate and lipid composition.

## 7.2. Physical Parameters

The size of native and desialylated LDL particles was measured by quasi-elastic laser scattering in a lipoprotein suspension and by electrophoresis in polyacrylamide gel followed by scanning densitometry [93]. The size distribution modes obtained for native LDL of healthy subjects and patients by the first method were 26.5 and 26.8 nm, respectively. The size distribution modes for desialylated LDL of healthy subjects and patients were 24.8 and

24.5 nm, respectively. Similar values were obtained by polyacrylamide gel electrophoresis analysis for average diameters of native LDL (26.3 and 26.2 nm) and desialylated LDL (23.5 and 22.9 nm) of healthy subjects and patients, respectively [92]. Thus, desialylated LDL particles are smaller than native LDL particles.

The density of native and desialylated LDL, was determined by gradient density ultracentrifugation [92]. It was shown that the distribution of desialylated LDL is shifted toward high densities in comparison with native LDL. Thus, desialylaed LDL is denser than native LDL. Increased density of desialylated LDL particles is associated with a decreased content of phospholipids, free and esterified cholesteryl, and triglycerides. Analysis of LDL subfractions showed that the increase of LDL density is accompanied with a smaller particle size, decreased content of sialic acid and higher atherogenicity. Therefore, the densest desialylated LDL possesses the highest ability to induce intracellular lipid accumulation.

Using agarose gel electrophoresis it was shown that desialylated LDL has 1.2- to 1.4-fold higher electrophoretic mobility than native LDL [92]. So, desialylated LDL is more electronegative than native LDL.

## 7.3. Analysis of Apoprotein B

In order to identify the modifications of apoprotein B of desialylated LDL, we have determined the content of free lysine groups in untreated and delipidated lipoprotein preparations [92]. The content of free amino groups was evaluated by color reaction with trinitrobenzenesulfonic acid. In healthy subjects, the contents of free amino groups in intact particles of native LDL and desialylated LDL did not differ significantly. Similar amounts of free amino groups were determined in delipidated apoB preparations of these lipoproteins.

The content of free amino groups in untreated desialylated LDL of patients was 2-fold lower than that of native LDL. The content of free amino groups in delipidated apoB of desialylated LDL was higher than in intact particles but lower than in delipidated apoB of native LDL. These findings indicate that part of amino groups of apoB is chemically modified, while another part is masked due to the changes in tertiary structure of apoB.

## 7.4. Degree of Oxidation and Oxidizability

We attempted to answer the following questions: 1) is circulating atherogenic desialylated LDL oxidized lipoprotein? 2) what is the reason for LDL susceptibility to oxidation? 3) is the degree of in vivo oxidation of LDL sufficient to render it atherogenicity?

Usually, the degree of lipoprotein oxidation is assessed by the contents of hydroperoxides or thiobarbituric acid-reactive substances (TBARS), compounds being formed upon lipid peroxidation. However, chemical instability and hydrophylic properties of these compounds may lead to its removal from LDL particle during lipoprotein isolation and purification. We have developed a new approach to the evaluation of the degree of LDL oxidation [97]. This approach is based on the hypothesis that chemically active lipid derivatives being formed during peroxidation covalently bind to apoprotein B, and thus may serve as a marker of lipoperoxidation occurring in vivo in lipoprotein particle. Using high-performance liquid chromatography and nuclear magnetic resonance, we have found sterols (predominantly cholesterol) and phosphates covalently bound to apoB in delipidated samples of LDL are oxidized by copper ions, azo-initiators, sodium hypochlorite or cultured cells. Freshly isolated LDL of healthy subjects contained no apoB-lipid adducts. It was shown that in contrast to other parameters used to evaluate the intensity of lipid peroxidation in LDL, the level of cholesterol covalently bound to apoB of copper-oxidized LDL rises monotonously during incubation [95]. Thus, the content of apoB-bound cholesterol in LDL is a parameter reflecting the degree of LDL oxidation.

The content of apoB-bound cholesterol in native LDL and desialylated LDL of healthy subjects was O.25±O.08 and 0.28±0.05 mol/mol apoB, respectively. The level of apoB-bound cholesterol in native LDL of patients with coronary atherosclerosis did not differ significantly from its level in native LDL of healthy subjects. The content of apoB-bound cholesterol in desialylated LDL of patients was 7-fold higher than in native LDL. Thus, we have shown that desialylated LDL of coronary atherosclerosis patients is oxidized lipoproteins.

Desialylated LDL contains 2-to 4-fold more oxysterols compared to native LDL [92]. These data can serve as proof of the higher oxidizability of desialylated LDL.

In addition to a high degree of in vivo oxidation, desialylated LDL possesses a higher susceptibility to in vitro oxidation. In vitro oxidizability of LDL was assessed by the duration of lag-phase upon oxidation by copper ions

[98]. The mean duration of lag-phase of native LDL from patients did not differ from that of native LDL from healthy subjects. The lag-phase of desialylated LDL of healthy subjects and patients was significantly shorter (3- and 6-fold, respectively) than that of native LDL, indicating a higher in vitro oxidizability of desialylated LDL. It should be noted that oxidizability of total LDL preparations from healthy subjects and patients positively correlates with the proportion of desialylated LDL in lipoprotein preparation.

In an attempt to find out the causes of increased degrees of in vivo oxidation and oxidizability of desialylated LDL, we determined the contents of the major fat-soluble antioxidants in lipoprotein particles, and analyzed the correlations between the contents of coenzyme-$Q_{10}$, tocopherols and carotenoids and the level of apoB-bound cholesterol and duration of lag-phase.

The content of all studied antioxidants (coenzyme-$Q_{10}$, α-and γ-tocopherols, β-carotene and lycopene) in desialylated LDL was 1.5- to 2-fold lower than in native LDL. The level of apoB-bound cholesterol in desialylated LDL, which reflects the degree of lipoprotein oxidation, positively correlated with the level of ubiquinone and showed negative correlation with ubiquinol and β-carotene levels. On the other hand, a positive correlation was established between the contents of apoB-bound cholesterol in native LDL and the ubiquinol level. The duration of lag-phase for desialylated LDL positively correlated with α-tocopherol and β-carorene contents and negatively correlated with the ubiquinone content. On the other hand, oxidizability of native LDL positively correlated with ubiquinone level.

From these data it can be concluded that: a) the content of all investigated lipid-soluble antioxidants in desialylated LDL is lower than in native lipoproteins; b) this fact determines the high oxidizability of desialylated LDL; c) coenzyme-$Q_{10}$ may act as a pro-oxidant in the native LDL subfraction; d) in vivo lipoperoxidation in desialylated LDL is corroborated by increased proportion of oxidized form of coenzyme-$Q_{10}$; e) the degree of desialylated LDL in vivo oxidation correlates with oxidation of ubiquinol and loss of carotenoids.

## 7.5. Immunogenicity (Anti-LDL Autoantibodies)

We have isolated total immunoglobulin G fraction from the sera of atherosclerotic patients, and immunoglobulins interacting with LDL (anti-LDL) were then purified by affinity chromatography on a sorbent with immobilized LDL [22]. From the sera of atherosclerotic patients, more anti-

LDL were isolated as compared to healthy individuals. We were not the first who found antibodies against LDL in human blood. The antibodies against LDL were found in the patients suffering from different diseases as well as in healthy subjects [99, 100]. It was demonstrated that immunoglobulins are major LDL-binding proteins in human plasma [101].

As mentioned above, autoantibodies against MDA-LDL have been found in the blood [19]. In our study, the affinity constant of anti-LDL to LDL obtained from the blood of healthy subjects was lower than to modified LDL. Table 2 shows the affinity constants of purified anti-LDL to different lipoproteins.

**Table 2. Affinity Constants of Lipoprotein-Anti-Low Density Lipoprotein Interaction[#]**

| Lipoprotein | Affinity constant $(\times 10^{-7} M^{-1})$ | LDL-induced Cholesterol increment (% over control) |
|---|---|---|
| LDL from healthy subjects | 2.4 | $2\pm10$ |
| Glycosylated LDL | 2.6 | $105 \pm14**$ |
| Acetylated LDL | 2.8 | $163\pm15*$ |
| Cu2+-oxidized LDL | 3.5 | $250\pm18*$ |
| Lipoprotein(a) | 3.6 | $102\pm7*$ |
| MDA-LDL | 10.9 | $305\pm21'$ |
| LDL from atherosclerotic patients | 11.3 | $246\pm31*$ |
| Desialylated LDL | 89.4 | $290\pm32*$ |

Data from one of three representative experiments are presented. The affinity constant of anti-LDL was determined using native and modified [$^{125}$I]LDL of known specific activity (200-1,000 cpm/ng apolipoprotein B). Known amounts of added radioactive and nonradioactive LDL and the amount of [$^{125}$I]LDL bound to anti-LDL were used to calculate the affinity constant.

*, Significant difference from LDL from healthy subjects (p<0.05).
[#] Adapted from [22], with permission.

The affinity constants for lipoprotein(a) obtained from the blood of a healthy donor as well as for glycosylated LDL, acetylated LDL, and LDL oxidized by Cu2+ were similar to that for LDL isolated from healthy subjects. The anti-LDL autoantibodies show higher affinity for LDL isolated from the blood of atherosclerotic patients as well as for MDA- LDL. The LDL desialylated in vitro by neuraminidase had the highest affinity constant among the modified LDL tested.

We can conclude that autoantibodies against LDL are produced in response to the appearance of desialylated LDL in the blood. Cross-reaction with the MDA-LDL can be explained by a similar conformation of certain epitopes of desialylated LDL and MDA-LDL.

Therefore, antibodies against MDA-LDL found Palinski and co-workers [19], and autoantibodies that we found in the blood of atherosclerotic patients [22] are the same antibodies against desialylated LDL but not against oxidized LDL. This finding calls into question one of the most important arguments of oxidative theory.

## 7.6. Relationship between Lipoprotein Modifications and LDL Atherogenicity

Naturally occurring atherogenic LDL circulating in the blood is small, dense, and more electronegative; LDL particles possess modified lipid, protein and carbohydrate moieties as compared to native LDL. Therefore, these lipoprotein particles can be regarded as MULTIPLE-MODIFIED LDL particles. To reveal what modifications determine LDL atherogenic potential, we have analyzed the correlations between changes in different chemical and physical parameters of LDL and LDL ability to induce lipid accumulation in SMA(+) cells of human aortic intima. A significant negative correlation ($r=-0.66$, $p<0.05$) was established between LDL atherogenicity and the sialic acid content.

Other parameters, such as size and the charge of LDL particle, the contents of phospholipids and neutral lipids, fat-soluble antioxidants and lipid peroxidation products, amount of free lysine amino groups, and the degree of oxidation and oxidizability of LDL, did not correlate with atherogenicity significantly [90, 95]. Thus, desialylation is probably the most important modification resulting in lipoprotein atherogenicity.

# 8. NATURALLY OCCURRING MULITIPLE-MODIFIED LDL AND OTHER KNOWN LDL MODIFICATIONS DETECTED IN HUMAN BLOOD PLASMA

As mentioned above, more electronegative LDL and small dense LDL were detected in human blood [52, 102]. We have carried out the comparative studies of in vivo modified LDLs. Cooperative study with group of Avogaro (Venice, Italy) showed that more electronegative LDL isolated by ion-exchange chromatography is desialylated LDL [103]. On the other hand, we demonstrated that our desialylated LDL subfraction is more electronegative [90, 95]. Both facts suggest that desialylated LDL and electronegative LDL subfractions are similar if not identical. We found that a particle of desialylated LDL is smaller and denser than that of native LDL, i.e., this LDL is a small dense lipoprotein. On the other hand, La-Belle and Krauss showed that small dense LDL has a low content of sialic acid, i.e., is desialylated [104]. These findings point to a similarity between the two types of modified LDL. Table 3 summarizes the data on chemical composition and physical parameters of the modified LDL detected in human blood. There is an obvious similarity between LDL characteristics. Therefore, we think that all subfractions of modified LDL isolated by different methods are represented by the same lipoprotein particles that had undergone multiple modifications.

**Table 3. Characteristics of Modified LDL Found in Human Blood Plasma**

| Parameter | Desialylated LDL | LDL(-) | Small dense LDL |
|---|---|---|---|
| Atherogenicity | ↑[87, 88] | ↑ [52] | ↑ [45, 89] |
| Size | ↓[90, 95] | ↓ [52] | ↓ [45, 102] |
| Density | ↑ [90, 95] | ? | ↑ [105, 106] |
| (-) Charge | ↑ [90] | ↑ [52] | ↑ [45] |
| Sialic acid | ↓ [86, 88] | ↓ [103, 104] | ↓ [49] |
| Cholesteryl esters | ↓ [90] | ↓[110] | ↓ [106, 107] |
| Phospholipids | ↓ [93] | ↓ [110] | ↓ [106, 107] |
| Protein/lipids | ↑ [90] | ↑ [110] | ↑ [105, 106] |
| Oxidizability | ↑ [89, 94] | ↑ [39] | ↑ [108, 109] |
| Antioxidants | ↓ [90, 95] | ↓ [110] | ↓ [108, 111] |
| Amino group modification | ↑ [90] | ↑ [39] | ? |
| Self-association | ↑ [66, 67] | ↑ [53] | ↑ [91] |

# 9. Multiple Modification of LDL in Blood Plasma

After revealing multiple modifications of LDL particles, it was important to reveal the mechanisms of this modification. It was suggested that cells of some organs can modify LDL or multiple modifiacations take place in blood plasma. A 24-h incubation of LDL at 37°C with intact endotheliocytes, hepatocytes, macrophages and smooth muscle cells, or with cell homogenates, did not change physical parameters and chemical properties of native LDL [48].

After incubation for 24 h at 37°C in whole blood or plasma obtained from atherosclerotic patients, the sialic acid content of LDL was 2-fold lower than that of LDL incubated with whole blood and plasma obtained from healthy individuals. Incubation with red and white blood cells had no effect on the sialic acid content. Thus, LDL modification takes place in blood plasma [48].

The following protocol was developed for detail I investigation of the processes of LDL modification [48]. Native LDL was isolated from plasma by ultracentrifugation followed by lectin chromatography. Plasma-derived serum free from apoB-containing lipoproteins was prepared by defibrination or remaining LDL-deficient plasma. LDL and serum were mixed in the same proportion as in the original plasma and incubated for different time intervals at 37°C.

After incubation, LDL was re-isolated by ultracentrifugation. This approach allowed us to exclude the effects of LDL formed from very low density and intermediate density lipoproteins during incubation. A decrease of the sialic acid content of initially native LDL was observed after 1 h of incubation with autologous plasma in parallel with the appearance of desialylated LDL, which was determined by lectin-sorbent assay (Table 4). While the LDL sialic acid level continuously decreased, LDL became capable of inducing the accumulation of total cholesterol in SMA(+) cells cultured from unaffected human aortic intima in as early as 3 hours of incubation. Starting from the 6th hour of LDL incubation with plasma, a monotonous decrease in phospholipid and neutral lipid content as well as LDL size was observed.

After 36 hours of incubation, an increase in the negative charge of lipoprotein particles was registered. Prolonged incubation of LDL with plasma-derived serum (48 and 72 hours) leads to the loss of α-tocopherol as well as to increase LDL susceptibility to oxidation and to accumulation of cholesterol covalently bound to apoB, a marker of lipoperoxidation. In parallel, degradation of apoB takes place.

**Table 4. Scheme of LDL modification** [#]

| 1 h | 3 h | 6 h | 12 h | 24 h | 36 h | 48 h |
|---|---|---|---|---|---|---|
| ↓ Sialic acid<br>↑ % of desialylated LDL | ↑ Atherogenicity | → Free cholesterol | → Size<br>→ Cholesteryl esters<br>→ Phospholipids | → Triglycerides | ↑ Electronegativity | ↑ apo B-bound cholesterol<br>↑ Susceptibility to oxidation<br>↑ Fluorescence<br>→ Vitamin E |

[#] Adapted from [48], with permission.

Thus, desialylation of LDL particles represents one of the first, or the primary act, of modification, which is apparently a sufficient prerequisite for the development of atherogenic properties. Subsequent modifications just enhance the atherogenic potential of LDL.

Therefore, multiple modification of LDL in plasma is a cascade of successive changes in the lipoprotein particle: desialylation, loss of lipids, reduction in the particle size, increase of its electronegative charge and peroxidation of lipids. These ideas fully explain the detection of various forms of LDL modification in the blood, namely: desialylated, small dense, electronegative, and oxidized. Apparently, in such a sequence, circulating LDL particles are transformed in the blood. It should be noted that, contrary to popular belief, the oxidative form of LDL modification is not the only form of modification but is also not the most important modification, because LDL oxidation occurs at the later stages of multiple modifications and does not significantly increase the atherogenic potential of multiply modified LDL.

## 10. TRANS-SIALYDASE

Since desialylation is the first atherogenic modification, or one of the first atherogenic modifications of the LDL particle, elucidation of the mechanism of LDL desialylation in blood is extremely important. It was found that more than 98% of sialic acid removed from LDL is observed in protein-bound fractions, but not in free form [48]. Thus, the enzyme removing sialic acid from LDL transfers it to other plasma acceptors, i.e., the enzyme is a trans-sialidase.

We have found that that trans-sialydase activity removes sialic acid not only from LDL, but also from other lipoproteins, glycoproteins and gangliosides. Removed sialic acid may be transferred to protein and lipid moiety of lipoprotein particles as well as to glycoproteins and sphingolipids of human serum. Gel-filtration chromatography of blood serum revealed trans-sialidase activity in lipoprotein fraction and in lipoprotein-deficient serum. Thus, trans-sialidase activity exists both in lipoproteins and in free form. The nature of the enzyme-to-lipoprotein linkage is not clear; we suggest electrostatic interaction.

We have isolated trans-sialidase (about 65 kDa) from lipoprotein deficient serum using affinity chromatography [48]. The enzyme levels in blood serum varies from 20 to 200 µg/ml. Trans-sialidase possesses three pH optima: 3.0, 5.0 and 7.0. Therefore, the enzyme may act both in blood and in cellular organelles with low pH levels (e.g. within lysosomes). Calcium and magnesium ions can increase trans-sialidase activity in vitro at millimolar concentrations. SH-groups are important for normal enzyme function. Blood components can serve as donors for trans-sialidase. An isolated enzyme removed sialic acid from LDL, IDL, VLDL, and HDL particles. The enzyme is also capable of transferring sialic acid from plasma glycoconjugates, namely proteins (fetuin, transferrin) and gangliosides (GM3, GD3, GM1, GD1a, GD1b) but the rate of sialic acid transfer from these glycoconjugates is much lower as compared to LDL. Sialylated LDL is a more preferable substrate than sialylated VLDL, IDL and especially HDL. The reason for this difference is unknown. It is important that decrease in the sialic acid transfer rate from apoB-containing lipoproteins correlates with increase of lipoprotein particle size. Probably the particle size determines trans-sialidase activity.

Treatment of native LDL with isolated trans-sialidase leads to desialylation of LDL and then to induce cholesteryl ester accumulation in human aortic intimal SMA(+) cells [48]. Thus, trans-sialidase may be involved in foam cell formation.

We do not know the physiological role of the human plasma trans-sialidase. Naturally, the enzyme may participate in processes depended on the sialylation and desialylation of different cellular and non-cellular components. Trans-sialidase is capable to modulate the activities of plasma enzymes, change the lifetime of glycoproteins, lipoproteins and cells, affect cell-to-cell interaction, etc. [112, 113].

Thus, we have found and characterized a key enzyme desialylating LDL. Trans-sialidase may play very important role in atherogenesis as a factor of atherogenic LDL modification. Trans-sialidase modifies lipoprotein interaction with arterial cells. LDL desialylated by trans-sialidase causes intracellular lipid accumulation with accompanied stimulation of proliferative activity and extracellular matrix synthesis. Therefore, trans-sialidase-induced LDL desialylation leads to all known cellular manifestations of atherosclerosis.

## 11. MECHANISMS POTENTIATING LDL ATHEROGENICITY

Theoretical calculations, based on experimental data on the rates of LDL uptake and degradation, indicate that in case of native LDL, a normal intimal cell would require up to 130 years to become a foam cell, providing that all cholesterol that would be taken by the cell remains inside the cell. In case of desialylated LDL, this time is shortened up to 15 years. Angiographic and ultrasonographic data show that atherosclerotic plaque occlusing half of the carotid artery lumen may be formed within several weeks or months. Obviously, the real rate of the foam cell formation is much higher than the calculated one. Therefore, it was reasonable to suggest that there are some mechanisms potentiating the atherogenicity of desialylated LDL. We have identified at least three of these mechanisms: self-association of LDL particles, formation of LDL-containing immune complexes and formation of LDL complexes with the connective tissue matrix components.

## 11.1. Association of Modified LDL

We have shown that in vivo and in vitro modified LDL particles spontaneously associate under the conditions of cell culture while native LDL particles do not associate [90, 95]. A positive correlation between atherogenicity of modified LDL and the degree of LDL association was established [66, 67]. Lipoprotein associates isolated by gel filtration induced a dramatic increase in the lipid accumulation in cultured human aortic intimal SMA(+) cells. Removal of LDL associates from the incubation medium by filtration through filters with pore diameter 0.1 μm completely prevented intracellular lipid accumulation. Thus, association increases atherogenic potential of LDL.

Experiments with iodinated lipoprotein associates showed that the uptake of associated LDL is 5- to 20-fold higher than that of non-associated LDL particles [65]. Cytochalasin B, an inhibitor of cellular phagocytosis, and latex beads, the object of phagocytosis, inhibit the uptake of LDL associates [65-67]. These findings indicate that LDL associates are taken up by phagocytosis. The rate of intracellular degradation of associated modified LDL apoB was 2- to 5-fold lower than the rate of degradation of apoB of non-associated particles. Thus, the high atherogenicity of lipoprotein associates results from enhanced uptake by phagocytosis and a low rate of intracellular degradation.

## 11.2. Formation of Circulating Immune Complexes

Multiple modifications of lipoprotein particles suggests appearance of antigens with subsequent production of autoantibodies against these antigens. In fact, we have found circulating immune complexes consisting of LDL and anti-LDL autoantibodies in blood of most coronary atherosclerosis patients [114-118].

A positive correlation between blood serum levels of LDL containing immune complexes, and severity of coronary and extra-coronary atherosclerosis has been demonstrated [117, 118].

We isolated LDL from circulating immune complexes by affinity chromatography on agarose with immobilized goat polyclonal antibodies against human LDL and studied its properties [116]. LDL from circulating immune complexes was desialylated, small, dense, more electronegative particles with low contents of neutral lipids and phospholipids, as well as neutral saccharides. The conformational changes in tertiary structure of apoB were also observed. Thus, LDL from circulating immune complexes is identical to desialylated LDL subfraction.

Using affinity chromatography on LDL-agarose we have isolated the autoantibodies to modified LDL from blood plasma of coronary atherosclerosis patients [22, 117]. These autoantibodies were identified as immunoglobulin G with an isoelectric point of about 8.5 (8.1-9.0), capable of interacting with the protein but not lipid moiety of LDL. As mentioned above, these autoantibodies interact with native, glycosylated, acetylated and oxidized LDL, displaying the highest affinity for malondialdehyde-treated LDL, desialylated LDL, and LDL of patients with coronary atherosclerosis.

The complexes of autoantibodies with native LDL stimulated lipid accumulation in SMA(+) cells cultured from uninvolved human aortic intima. Moreover, autoantibodies potentiated atherogenic potential of desialylated LDL by forming complexes with lipoprotein [22, 117]. The binding of C1q component of complement and fibronectin to the LDL-autoantibody complex resulted in a more pronounced lipid accumulation in human aortic intimal SMA(+) cells. C1q component of complement is known to be expressed by interdigitating and follicular dendritic cells in the spleen, where C1q is thought to be involved in capturing immune complexes [119]. Antigen-presenting dendritic cells also reside in atherosclerotic lesions [120, 121]. In atherosclerotic lesions, dendritic cells expressing C1q were detected [122]. Apart from the identification of lesional C1q(+) dendritic cells, C1q expression was found in macrophages, macrophage foam cells and in neovascular endothelial cells [122]. It has been concluded that the expression of C1q, by cells residing in the arterial wall, might be important in binding and trapping immune complexes in atherosclerotic lesions [122].

The *in vitro* interaction of mouse peritoneal and human pericardial macrophages with immune complexes, which contained lipoprotein-antibody and which were isolated from the serum of ischemic heart disease (IHD) patients, led to the transformation of macrophages into foam cells [100]. Electron-microscopic analysis of macrophages incubated *in vitro* for 3 days with immune complexes, containing lipoprotein-antibody, showed that the cytoplasm of macrophages contained lipid vacuoles and that cisterns of endoplasmic reticulum (ER) were dramatically enlarged, and were filled with lipids (Figure 1). The accumulation of lipids within ER cisterns in macrophages might be accompanied by ER stress, which plays a non-redundant role in the pathogenesis of atherosclerosis [123].

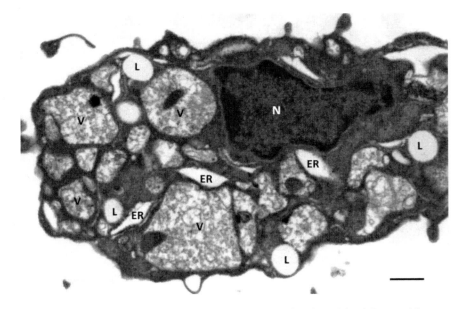

Figure 1. Ultrastructural Appearance of a Macrophage incubated for 3 Days with Immune Complexes, containing Lipoprotein-Antibody.
N – nucleus. V- vacuole.
L – lipid inclusion ("Lipid droplet"). Arrows show ER - endoplasmic reticulum.
Note that cisterns of ER are enlarged and are filled with lipids. Scale bar = 2 µm.

## 11.3. LDL Complexes with the Extracellular Matrix Components

We have also demonstrated that LDL can form complexes with cellular debris, collagen, elastin, and proteoglycans of human aortic intima [124, 125]. Addition of these complexes to cultured cells stimulated intracellular accumulation of lipids. Experiments with iodinated LDL showed an increased uptake and decreased intracellular degradation of lipoproteins in complexes compared to individual lipoprotein particles.

Thus, formation of large complexes containing desialylated LDL (self-associates, immune complexes and complexes with the connective tissue matrix) markedly increases atherogenic potential of modified lipoproteins.

## CONCLUSION

We have found a subfraction of LDL that is capable of inducing accumulation of lipids, primarily cholesteryl esters, in SMA(+) cells derived from undiseased human aortic intima. Thus, the paradoxical situation when native LDL induces no deposition of intracellular fat, and in vitro modified LDL is absent in circulation, has been resolved.

We have shown that atherogenic LDL is characterized by numerous alterations of carbohydrate, protein, and lipid moieties, and can be termed multiple-modified LDL. Multiple modifications of LDL occurs in human blood plasma. It was shown that circulating multiple-modified LDL loses the affinity for the B, E-receptor, and acquires the ability to interact with a number of other cellular membrane receptors and proteoglycans. The enhanced cellular uptake of desialylated LDL, low degradation rate of apolipoprotein and cholesteryl esters, as well as the stimulation of re-esterification of free cholesterol, cause the intracellular accumulation of intracellular esterified cholesterol.

The formation of LDL-containing large complexes (associates, immune complexes, complexes with the extracellular matrix components) can stimulate lipid accumulation in intimal smooth muscle cells. In addition to cholesteryl ester accumulation, desialylated LDL stimulates cell proliferation and synthesis of the connective tissue matrix.

Thus, we have discovered and characterized naturally occurring multiple-modified LDL capable of inducing all atherosclerotic manifestations at the cellular level.

## ACKNOWLEDGMENTS

This work was supported by the Russian Scientific Foundation (Grant # 14-15-00112), Russia.

## Conflict of Interest

The authors declare no conflict of interest.

## REFERENCES

[1]    Matsuura E, Atzeni F, Sarzi-Puttini P, Turiel M, Lopez LR, Nurmohamed MT. Is atherosclerosis an autoimmune disease? *BMC Med.* 2014, 12:47.

[2]    Reynolds T. Cholesteryl ester storage disease: a rare and possibly treatable cause of premature vascular disease and cirrhosis. *J. Clin. Pathol.* 2013;66(11):918-23.

[3]    Nicolaou G; Erridge C. Toll-like receptor-dependent lipid body formation in macrophage foam cell formation. *Curr. Opin. Lipidol.* 2010;21(5):427-33.

[4]    Stephen SL, Freestone K, Dunn S, Twigg MW, Homer-Vanniasinkam S, Walker JH, Wheatcroft SB, Ponnambalam S. Scavenger receptors and their potential as therapeutic targets in the treatment of cardiovascular disease. *Int. J. Hypertens.* 2010;2010:646929.

[5]    Yu XH, Zhang J, Zheng XL, Yang YH, Tang CK. Interferon-γ in foam cell formation and progression of atherosclerosis. *Clin. Chim. Acta.* 2015;441C:33-43.

[6]    Uitz E, Bahadori B, McCarty MF, Moghadasian MH. Practical strategies for modulating foam cell formation and behavior. *World J. Clin. Cases.* 2014;2(10):497-506.

[7]    Chen RM, Fischer-Dzoga K. Effect of hyperlipemic serum lipoproteins on the lipid accumulation and cholesterol flux of rabbit aortic medial cells. *Atherosclerosis.* 1977;28(3):339-53.

[8]    Goldstein JL, Anderson RG, Brown MS. Coated pits, coated vesicles, and receptor-mediated endocytosis. *Nature.* 1979;279(5715):679-85.

[9]    Coetzee GA, Stein O, Stein Y. Modulation by sodium ascorbate of the effect of chloroquine on low density lipoprotein retention and degradation in cultured human skin fibroblasts. *Atherosclerosis.* 1979;32(3):277-87.

[10]   Brown MS, Goldstein JL, Krieger M, Ho YK, Anderson RG. Reversible accumulation of cholesteryl esters in macrophages incubated with acetylated lipoproteins. *J. Cell Biol.* 1979;82(3):597-613.

[11]   Goldstein JL, Ho YK, Basu SK, Brown MS. Binding site on macrophages that mediates uptake and degradation of acetylated low density lipoprotein, producing massive cholesterol deposition. *Proc. Natl. Acad. Sci. U S A.* 1979;76(1):333-7.

[12]   Fogelman AM, Shechter I, Seager J, Hokom M, Child JS, Edwards PA. Malondialdehyde alteration of low density lipoproteins leads to cholesteryl ester accumulation in human monocyte-macrophages. *Proc. Natl. Acad. Sci. U S A.* 1980;77(4):2214-8.

[13]   Steinbrecher UP, Parthasarathy S, Leake DS, Witztum JL, Steinberg D. Modification of low density lipoprotein by endothelial cells involves lipid peroxidation and degradation of low density lipoprotein phospholipids. *Proc. Natl. Acad. Sci. U S A.* 1984;81(12):3883-7.

[14]   Haberland ME, Olch CL, Folgelman AM. Role of lysines in mediating interaction of modified low density lipoproteins with the scavenger receptor of human monocyte macrophages. *J. Biol. Chem.* 1984;259(18):11305-11.

[15]   Prescott MF, Müller KR, Flammer R, Feige U. The effect of LDL and modified LDL on macrophage secretion products. *Agents Actions Suppl.* 1984;16:163-70.

[16]   Trpkovic A, Resanovic I, Stanimirovic J, Radak D, Mousa SA, Cenic-Milosevic D, Jevremovic D, Isenovic ER. Oxidized low-density lipoprotein as a biomarker of cardiovascular diseases. *Crit. Rev. Clin. Lab Sci.* 2014; 24:1-16.

[17]   Arai H. Oxidative modification of lipoproteins. *Subcell Biochem.* 2014;77:103-14.

[18]   Steinberg D, Witztum JL. Oxidized low-density lipoprotein and atherosclerosis. *Arterioscler Thromb Vasc Biol.* 2010;30(12):2311-6.

[19]   Palinski W, Rosenfeld ME, Ylä-Herttuala S, Gurtner GC, Socher SS, Butler SW, Parthasarathy S, Carew TE, Steinberg D, Witztum JL. Low density lipoprotein undergoes oxidative modification in vivo. *Proc. Natl. Acad. Sci. U S A.* 1989 Feb;86(4):1372-6.

[20]  Fukuchi M, Watanabe J, Kumagai K, Baba S, Shinozaki T, Miura M, Kagaya Y, Shirato K. Normal and oxidized low density lipoproteins accumulate deep in physiologically thickened intima of human coronary arteries. *Lab. Invest.* 2002;82(10):1437-47.

[21]  Ylä-Herttuala S, Palinski W, Rosenfeld ME, Steinberg D, Witztum JL. Lipoproteins in normal and atherosclerotic aorta. *Eur. Heart J.* 1990;11 Suppl E:88-99.

[22]  Orekhov AN, Tertov VV, Kabakov AE, Adamova IYu, Pokrovsky SN, Smirnov VN. Autoantibodies against modified low density lipoprotein. Nonlipid factor of blood plasma that stimulates foam cell formation. *Arterioscler. Thromb.* 1991;11(2):316-26.

[23]  Mills EJ, Rachlis B, Wu P, Devereaux PJ, Arora P, Perri D. Primary prevention of cardiovascular mortality and events with statin treatments: a network meta-analysis involving more than 65,000 patients. *J. Am. Coll. Cardiol.* 2008;52(22):1769-81.

[24]  Berneis K, Jeanneret C, Muser J, Felix B, Miserez AR. Low-density lipoprotein size, and subclasses are markers of clinically apparent and non-apparent atherosclerosis in type 2 diabetes. *Metabolism.* 2005 Feb;54(2):227-34.

[25]  Krauss RM, Lindgren FT, Ray RM. Interrelationships among subgroups of serum lipoproteins in normal human subjects. *Clin. Chim. Acta.* 1980;104(3):275-90.

[26]  Havel RJ, Eder HA, Bragdon JH. The distribution and chemical composition of ultracentrifugally separated lipoproteins in human serum. *J. Clin. Invest.* 1955;34(9):1345-53.

[27]  Gofman JW, Lindgren FT, Elliott H. Ultracentrifugal studies of lipoproteins of human serum. *The Journal of biological chemistry* 1949; 179: 973-979.

[28]  Ensign W, Hill N, Heward CB. Disparate LDL phenotypic classification among 4 different methods assessing LDL particle characteristics. *Clinical chemistry* 2006; 52: 1722-1727.

[29]  Williams PT, Vranizan KM, Krauss RM. Correlations of plasma lipoproteins with LDL subfractions by particle size in men and women. *Journal of lipid research* 1992; 33: 765-774.

[30]  Hoefner DM, Hodel SD, O'Brien JF, Branum EL, Sun D, Meissner I, McConnell JP. Development of a rapid, quantitative method for LDL subfractionation with use of the Quantimetrix Lipoprint LDL System. *Clin. Chem.* 2001;47(2):266-74.

[31] Otvos JD, Jeyarajah EJ, Bennett DW, Krauss RM. Development of a proton nuclear magnetic resonance spectroscopic method for determining plasma lipoprotein concentrations and subspecies distributions from a single, rapid measurement. *Clin. Chem.* 1992;38(9):1632-8.

[32] Witte DR, Taskinen MR, Perttunen-Nio H, Van Tol A, Livingstone S, Colhoun HM. Study of agreement between LDL size, as measured by nuclear magnetic resonance and gradient gel electrophoresis. *J. Lipid. Res.* 2004;45(6):1069-76.

[33] Okazaki M, Usui S, Ishigami M, Sakai N, Nakamura T, Matsuzawa Y, Yamashita S. Identification of unique lipoprotein subclasses for visceral obesity by component analysis of cholesterol profile in high-performance liquid chromatography. *Arterioscler. Thromb. Vasc. Biol.* 2005;25(3):578-84.

[34] O'Neal D, Harrip P, Dragicevic G, Rae D, Best JD. A comparison of LDL size determination using gradient gel electrophoresis and light-scattering methods. *J. Lipid. Res.* 1998;39(10):2086-90.

[35] Sakurai T, Trirongjitmoah S, Nishibata Y, Namita T, Tsuji M, Hui SP, Jin S, Shimizu K, Chiba H. Measurement of lipoprotein particle sizes using dynamic light scattering. *Ann. Clin. Biochem.* 2010 Sep;47(Pt 5):476-81.

[36] Caulfield MP, Li S, Lee G, Blanche PJ, Salameh WA, Benner WH, Reitz RE, Krauss RM. Direct determination of lipoprotein particle sizes and concentrations by ion mobility analysis. *Clin. Chem.* 2008;54(8):1307-16.

[37] Musunuru K, Orho-Melander M, Caulfield MP, Li S, Salameh WA, Reitz RE, Berglund G, Hedblad B, Engström G, Williams PT, Kathiresan S, Melander O, Krauss RM. Ion mobility analysis of lipoprotein subfractions identifies three independent axes of cardiovascular risk. *Arterioscler. Thromb. Vasc. Biol.* 2009;29(11):1975-80.

[38] Hirano T, Ito Y, Saegusa H, Yoshino G. A novel and simple method for quantification of small, dense LDL. *J. Lipid Res.* 2003;44(11):2193-201.

[39] Berneis KK, Krauss RM. Metabolic origins and clinical significance of LDL heterogeneity. Journal of lipid research 2002; 43: 1363-1379.

[40] Krauss RM, Williams PT, Lindgren FT, Wood PD. Coordinate changes in levels of human serum low and high density lipoprotein subclasses in healthy men. *Arteriosclerosis.* 1988;8(2):155-62.

[41] Musunuru K, Strong A, Frank-Kamenetsky M, Lee NE, Ahfeldt T, Sachs KV, Li X, Li H, Kuperwasser N, Ruda VM, Pirruccello JP, Muchmore B, Prokunina-Olsson L, Hall JL, Schadt EE, Morales CR, Lund-Katz S, Phillips MC, Wong J, Cantley W, Racie T, Ejebe KG, Orho-Melander M, Melander O, Koteliansky V, Fitzgerald K, Krauss RM, Cowan CA, Kathiresan S, Rader DJ. From noncoding variant to phenotype via SORT1 at the 1p13 cholesterol locus. *Nature.* 2010;466(7307):714-9.

[42] Packard C, Caslake M, Shepherd J. The role of small, dense low density lipoprotein (LDL): a new look. *Int. J. Cardiol.* 2000;74 Suppl 1:S17-22.

[43] Griffin BA. Lipoprotein atherogenicity: an overview of current mechanisms. Proc. Nutr. Soc. 1999 Feb;58(1):163-9.

[44] Steinberg D, Parthasarathy S, Carew TE, Khoo JC, Witztum JL. Beyond cholesterol. Modifications of low-density lipoprotein that increase its atherogenicity. *N. Engl. J. Med.* 1989;320(14):915-24.

[45] Jaakkola O, Solakivi T, Tertov VV, Orekhov AN, Miettinen TA, Nikkari T. Characteristics of low-density lipoprotein subfractions from patients with coronary artery disease. *Coron Artery Dis.* 1993 Apr;4(4):379-85.

[46] Tribble DL, Rizzo M, Chait A, Lewis DM, Blanche PJ, Krauss RM. Enhanced oxidative susceptibility and reduced antioxidant content of metabolic precursors of small, dense low-density lipoproteins. *Am. J. Med.* 2001;110(2):103-10.

[47] Tertov VV, Kaplun VV, Sobenin IA, Boytsova EY, Bovin NV, Orekhov AN. Human plasma trans-sialidase causes atherogenic modification of low density lipoprotein. *Atherosclerosis.* 2001;159(1):103-15.

[48] Tertov VV, Kaplun VV, Sobenin IA, Orekhov AN. Low-density lipoprotein modification occurring in human plasma possible mechanism of in vivo lipoprotein desialylation as a primary step of atherogenic modification. *Atherosclerosis.* 1998;138(1):183-95.

[49] La Belle M, Krauss RM. Differences in carbohydrate content of low density lipoproteins associated with low density lipoprotein subclass patterns. *J. Lipid. Res.* 1990;31(9):1577-88.

[50] Anber V, Griffin BA, McConnell M, Packard CJ, Shepherd J. Influence of plasma lipid and LDL-subfraction profile on the interaction between low density lipoprotein with human arterial wall proteoglycans. *Atherosclerosis.* 1996;124(2):261-71.

[51] Hoff HF, Gaubatz JW. Isolation, purification, and characterization of a lipoprotein containing Apo B from the human aorta. *Atherosclerosis.* 1982;42(2-3):273-97.

[52]  Avogaro P, Bon GB, Cazzolato G. Presence of a modified low density lipoprotein in humans. *Arteriosclerosis* 1988; 8:79-87.

[53]  Chen CH, Jiang T, Yang JH, Jiang W, Lu J, Marathe GK, Pownall HJ, Ballantyne CM, McIntyre TM, Henry PD, Yang CY. Low-density lipoprotein in hypercholesterolemic human plasma induces vascular endothelial cell apoptosis by inhibiting fibroblast growth factor 2 transcription. *Circulation.* 2003;107(16):2102-8.

[54]  Yang CY, Raya JL, Chen HH, Chen CH, Abe Y, Pownall HJ, Taylor AA, Smith CV. Isolation, characterization, and functional assessment of oxidatively modified subfractions of circulating low-density lipoproteins. *Arterioscler. Thromb. Vasc. Biol.* 2003;23(6):1083-90.

[55]  Tang D, Lu J, Walterscheid JP, Chen HH, Engler DA, Sawamura T, Chang PY, Safi HJ, Yang CY, Chen CH. Electronegative LDL circulating in smokers impairs endothelial progenitor cell differentiation by inhibiting Akt phosphorylation via LOX-1. *J. Lipid. Res.* 2008 ;49(1):33-47.

[56]  Chan HC, Ke LY, Chu CS, Lee AS, Shen MY, Cruz MA, Hsu JF, Cheng KH, Chan HC, Lu J, Lai WT, Sawamura T, Sheu SH, Yen JH, Chen CH. Highly electronegative LDL from patients with ST-elevation myocardial infarction triggers platelet activation and aggregation. *Blood.* 2013;122(22):3632-41.

[57]  Chang PY, Chen YJ, Chang FH, Lu J, Huang WH, Yang TC, Lee YT, Chang SF, Lu SC, Chen CH. Aspirin protects human coronary artery endothelial cells against atherogenic electronegative LDL via an epigenetic mechanism: a novel cytoprotective role of aspirin in acute myocardial infarction. *Cardiovasc. Res.* 2013;99(1):137-45.

[58]  Schmitz G, Mollers C, Richter V. Analytical capillary isotachophoresis of human serum lipoproteins. *Electrophoresis* 1997; 18: 1807-1813.

[59]  Zhang B, Kaneshi T, Ohta T, Saku K. Relation between insulin resistance and fast-migrating LDL subfraction as characterized by capillary isotachophoresis. *J. Lipid Res.* 2005;46(10):2265-77.

[60]  Mello AP, da Silva IT, Abdalla DS, Damasceno NR. Electronegative low-density lipoprotein: origin and impact on health and disease. *Atherosclerosis.* 2011;215(2):257-65.

[61]  Damasceno NR, Sevanian A, Apolinário E, Oliveira JM, Fernandes I, Abdalla DS. Detection of electronegative low density lipoprotein (LDL-) in plasma and atherosclerotic lesions by monoclonal antibody-based immunoassays. *Clin. Biochem.* 2006;39(1):28-38.

[62] Santo Faulin Tdo E, de Sena KC, Rodrigues Telles AE, de Mattos Grosso D, Bernardi Faulin EJ, Parra Abdalla DS. Validation of a novel ELISA for measurement of electronegative low-density lipoprotein. *Clin. Chem. Lab. Med.* 2008;46(12):1769-75.

[63] Faulin Tdo E, de Sena-Evangelista KC, Pacheco DB, et al. Development of immunoassays for anti-electronegative LDL autoantibodies and immune complexes. Clinica chimica acta; *international journal of clinical chemistry* 2012; 413: 291-297.

[64] Parasassi T, De Spirito M, Mei G, Brunelli R, Greco G, Lenzi L, Maulucci G, Nicolai E, Papi M, Arcovito G, Tosatto SC, Ursini F. Low density lipoprotein misfolding and amyloidogenesis. *FASEB J.* 2008;22(7):2350-6.

[65] Tertov VV, Sobenin IA, Gabbasov ZA, Popov EG, Yaroslavov AA, Jauhiainen M, Ehnholm C, Smirnov VN, Orekhov AN. Three types of naturally occurring modified lipoproteins induce intracellular lipid accumulation in human aortic intimal cells--the role of lipoprotein aggregation. *Eur. J. Clin. Chem. Clin. Biochem.* 1992;30(4):300-8.

[66] Tertov VV, Sobenin IA, Gabbasov ZA, Popov EG, Orekhov AN. Lipoprotein aggregation as an essential condition of intracellular lipid accumulation caused by modified low density lipoproteins. *Biochem. Biophys. Res. Commun.* 1989;163(1):489-94.

[67] Tertov VV, Orekhov AN, Sobenin IA, Gabbasov ZA, Popov EG, Yaroslavov AA, Smirnov VN. Three types of naturally occurring modified lipoproteins induce intracellular lipid accumulation due to lipoprotein aggregation. *Circ. Res.* 1992;71(1):218-28.

[68] Aksenov DV, Medvedeva LA, Skalbe TA, Sobenin IA, Tertov VV, Gabbasov ZA, Popov EV, Orekhov AN. Deglycosylation of apo B-containing lipoproteins increase their ability to aggregate and to promote intracellular cholesterol accumulation in vitro. *Arch. Physiol. Biochem.* 2008;114(5):349-56.

[69] Brunelli R, Balogh G, Costa G, De Spirito M, Greco G, Mei G, Nicolai E, Vigh L, Ursini F, Parasassi T. Estradiol binding prevents ApoB-100 misfolding in electronegative LDL(-). *Biochemistry.* 2010;49(34):7297-302.

[70] Parasassi T, Bittolo-Bon G, Brunelli R, Cazzolato G, Krasnowska EK, Mei G, Sevanian A, Ursini F. Loss of apoB-100 secondary structure and conformation in hydroperoxide rich, electronegative LDL(-). *Free Radic. Biol. Med.* 2001;31(1):82-9.

[71] Blanco FJ, Villegas S, Benítez S, Bancells C, Diercks T, Ordóñez-Llanos J, Sánchez-Quesada JL. 2D-NMR reveals different populations of exposed lysine residues in the apoB-100 protein of electronegative and electropositive fractions of LDL particles. *J. Lipid. Res.* 2010 Jun;51(6):1560-5.

[72] Benítez S, Sánchez-Quesada JL, Lucero L, Arcelus R, Ribas V, Jorba O, Castellví A, Alonso E, Blanco-Vaca F, Ordóñez-Llanos J. Changes in low-density lipoprotein electronegativity and oxidizability after aerobic exercise are related to the increase in associated non-esterified fatty acids. *Atherosclerosis.* 2002;160(1):223-32.

[73] Gaubatz JW, Gillard BK, Massey JB, Hoogeveen RC, Huang M, Lloyd EE, Raya JL, Yang CY, Pownall HJ. Dynamics of dense electronegative low density lipoproteins and their preferential association with lipoprotein phospholipase A(2). *J. Lipid Res.* 2007;48(2):348-57.

[74] Jayaraman S, Gantz DL, Gursky O. Effects of phospholipase A(2) and its products on structural stability of human LDL: relevance to formation of LDL-derived lipid droplets. *Journal of lipid research* 2011; 52: 549-557.

[75] Benítez S, Villegas V, Bancells C, Jorba O, González-Sastre F, Ordóñez-Llanos J, Sánchez-Quesada JL. Impaired binding affinity of electronegative low-density lipoprotein (LDL) to the LDL receptor is related to nonesterified fatty acids and lysophosphatidyl choline content. *Biochemistry.* 2004;43(50):15863-72.

[76] Urata J, Ikeda S, Koga S, Nakata T, Yasunaga T, Sonoda K, Koide Y, Ashizawa N, Kohno S, Maemura K. Negatively charged low-density lipoprotein is associated with atherogenic risk in hypertensive patients. *Heart Vessels.* 2012;27(3):235-42.

[77] Lu J, Yang JH, Burns AR, Chen HH, Tang D, Walterscheid JP, Suzuki S, Yang CY, Sawamura T, Chen CH. Mediation of electronegative low-density lipoprotein signaling by LOX-1: a possible mechanism of endothelial apoptosis. *Circ. Res.* 2009 Mar 13;104(5):619-27.

[78] Chu CS, Wang YC, Lu LS, Walton B, Yilmaz HR, Huang RY, Sawamura T, Dixon RA, Lai WT, Chen CH, Lu J. Electronegative low-density lipoprotein increases C-reactive protein expression in vascular endothelial cells through the LOX-1 receptor. *PLoS One.* 2013;8(8):e70533.

[79] Greco G, Balogh G, Brunelli R, Costa G, De Spirito M, Lenzi L, Mei G, Ursini F, Parasassi T. Generation in human plasma of misfolded,

aggregation-prone electronegative low density lipoprotein. *Biophys. J.* 2009;97(2):628-35.

[80]  Sánchez-Quesada JL, Camacho M, Antón R, Benítez S, Vila L, Ordóñez-Llanos J. Electronegative LDL of FH subjects: chemical characterization and induction of chemokine release from human endothelial cells. *Atherosclerosis.* 2003;166(2):261-70.

[81]  Benítez S, Camacho M, Bancells C, Vila L, Sánchez-Quesada JL, Ordóñez-Llanos J. Wide proinflammatory effect of electronegative low-density lipoprotein on human endothelial cells assayed by a protein array. *Biochim. Biophys. Acta.* 2006;1761(9):1014-21.

[82]  Orekhov AN, Tertov VV, Novikov ID, Krushinsky AV, Andreeva ER, Lankin VZ, Smirnov VN. Lipids in cells of atherosclerotic and uninvolved human aorta. I. Lipid composition of aortic tissue and enzyme-isolated and cultured cells. *Exp. Mol. Pathol.* 1985;42(1):117-37.

[83]  Orekhov AN, Tertov VV, Pokrovsky SN, Adamova IYu, Martsenyuk ON, Lyakishev AA, Smirnov VN. Blood serum atherogenicity associated with coronary atherosclerosis. Evidence for nonlipid factor providing atherogenicity of low-density lipoproteins and an approach to its elimination. *Circ. Res.* 1988;62(3):421-9.

[84]  Tertov VV, Orekhov AN, Martsenyuk ON, Perova NV, Smirnov VN. Low-density lipoproteins isolated from the blood of patients with coronary heart disease induce the accumulation of lipids in human aortic cells. *Exp. Mol. Pathol.* 1989;50(3):337-47.

[85]  Chazov EI, Tertov VV, Orekhov AN, Lyakishev AA, Perova NV, Kurdanov KA, Khashimov KA, Novikov ID, Smirnov VN. Atherogenicity of blood serum from patients with coronary heart disease. *Lancet.* 1986;2(8507):595-8.

[86]  Orekhov AN, Tertov VV, Mukhin DN. Desialylated low density lipoprotein--naturally occurring modified lipoprotein with atherogenic potency. *Atherosclerosis.* 1991;86(2-3):153-61.

[87]  Orekhov AN, Tertov VV, Mukhin DN, Mikhailenko IA. Modification of low density lipoprotein by desialylation causes lipid accumulation in cultured cells: discovery of desialylated lipoprotein with altered cellular metabolism in the blood of atherosclerotic patients. *Biochem. Biophys Res. Commun.* 1989;162(1):206-11.

[88]  Tertov VV, Sobenin IA, Tonevitsky AG, Orekhov AN, Smirnov VN. Isolation of atherogenic modified (desialylated) low density lipoprotein from blood of atherosclerotic patients: separation from native lipoprotein

by affinity chromatography. *Biochem. Biophys. Res. Commun.* 1990;167(3):1122-7.

[89]   Tertov VV, Sobenin IA, Orekhov AN. Modified (desialylated) low-density lipoprotein measured in serum by lectin-sorbent assay. *Clin. Chem.* 1995 Jul;41(7):1018-21.

[90]   Tertov VV, Sobenin IA, Gabbasov ZA, Popov EG, Jaakkola O, Solakivi T, Nikkari T, Smirnov VN, Orekhov AN. Multiple-modified desialylated low density lipoproteins that cause intracellular lipid accumulation. Isolation, fractionation and characterization. *Lab. Invest.* 1992;67(5):665-75.

[91]   Orekhov AN, Tertov VV, Sobenin IA, Smirnov VN, Via DP, Guevara J Jr, Gotto AM Jr, Morrisett JD. Sialic acid content of human low density lipoproteins affects their interaction with cell receptors and intracellular lipid accumulation. *J. Lipid Res.* 1992;33(6):805-17.

[92]   Tertov VV, Orekhov AN. Metabolism of native and naturally occurring multiple modified low density lipoprotein in smooth muscle cells of human aortic intima. *Exp. Mol. Pathol.* 1997;64(3):127-45.

[93]   Orekhov AN, Tertov VV, Kudryashov SA, Smirnov VN. Triggerlike stimulation of cholesterol accumulation and DNA and extracellular matrix synthesis induced by atherogenic serum or low density lipoprotein in cultured cells. *Circ. Res.* 1990 Feb;66(2):311-20.

[94]   Tertov VV, Orekhov AN, Ryong LH, Smirnov VN. Intracellular cholesterol accumulation is accompanied by enhanced proliferative activity of human aortic intimal cells. *Tissue Cell.* 1988;20(6):849-54.

[95]   Tertov VV, Sobenin IA, Orekhov AN. Characterization of desialylated low-density lipoproteins which cause intracellular lipid accumulation. *Int. J. Tissue React.* 1992;14(4):155-62.

[96]   Tertov VV, Orekhov AN, Sobenin IA, Morrisett JD, Gotto AM Jr, Guevara JG Jr. Carbohydrate composition of protein and lipid components in sialic acid-rich and -poor low density lipoproteins from subjects with and without coronary artery disease. *J. Lipid Res.* 1993;34(3):365-75.

[97]   Tertov VV, Kaplun VV, Dvoryantsev SN, Orekhov AN. Apolipoprotein B-bound lipids as a marker for evaluation of low density lipoprotein oxidation in vivo. *Biochem. Biophys. Res. Commun.* 1995;214(2):608-13.

[98]   Esterbauer H, Striegl G, Puhl H, Rotheneder M. Continuous monitoring of in vitro oxidation of human low density lipoprotein. *Free Radic. Res. Commun.* 1989;6(1):67-75.

[99] Beaumont JL. Auto-immune hyperlipidemia. An atherogenic metabolic disease of immune origin. *Rev. Eur. Etud. Clin. Biol.* 1970;15(10):1037-41.

[100] Klimov AN, Denisenko AD, Vinogradov AG, Nagornev VA, Pivovarova YI, Sitnikova OD, Pleskov VM. Accumulation of cholesteryl esters in macrophages incubated with human lipoprotein-antibody autoimmune complex. *Atherosclerosis.* 1988;74(1-2):41-6.

[101] Bauer BJ, Blashfield K, Norris R, Buthala DA, Ginsberg LC. Immunoglobulin as the major low density lipoprotein binding protein in plasma. *Atherosclerosis.* 1982;44(2):153-60.

[102] Krauss RM, Burke DJ. Identification of multiple subclasses of plasma low density lipoproteins in normal humans. *J. Lipid Res.* 1982;23(1):97-104.

[103] Tertov VV, Bittolo-Bon G, Sobenin IA, Cazzolato G, Orekhov AN, Avogaro P. Naturally occurring modified low density lipoproteins are similar if not identical: more electronegative and desialylated lipoprotein subfractions. *Exp. Mol. Pathol.* 1995;62(3):166-72.

[104] Tertov VV, Sobenin IA, Orekhov AN. Similarity between naturally occurring modified desialylated, electronegative and aortic low density lipoprotein. *Free Radic. Res.* 1996;25(4):313-9.

[105] Lindgren FT, Jensen LC, Wills RD, Freeman NK. Flotation rates, molecular weights and hydrated densities of the low-density lipoproteins. *Lipids.* 1969;4(5):337-44.

[106] Shen MM, Krauss RM, Lindgren FT, Forte TM. Heterogeneity of serum low density lipoproteins in normal human subjects. *J. Lipid Res.* 1981 Feb;22(2):236-44.

[107] Teng B, Thompson GR, Sniderman AD, Forte TM, Krauss RM, Kwiterovich PO Jr. Composition and distribution of low density lipoprotein fractions in hyperapobetalipoproteinemia, normolipidemia, and familial hypercholesterolemia. *Proc. Natl. Acad. Sci. U S A.* 1983;80(21):6662-6.

[108] Dejager S, Bruckert E, Chapman MJ. Dense low density lipoprotein subspecies with diminished oxidative resistance predominate in combined hyperlipidemia. *J. Lipid Res.* 1993;34(2):295-308.

[109] Tribble DL, van den Berg JJ, Motchnik PA, Ames BN, Lewis DM, Chait A, Krauss RM. Oxidative susceptibility of low density lipoprotein subfractions is related to their ubiquinol-10 and alpha-tocopherol content. *Proc. Natl. Acad. Sci. U S A.* 1994;91(3):1183-7.

[110] Avogaro P, Cazzolato G, Bittolo-Bon G. Some questions concerning a small, more electronegative LDL circulating in human plasma. *Atherosclerosis.* 1991;91(1-2):163-71.

[111] Cazzolato G, Avogaro P, Bittolo-Bon G. Characterization of a more electronegatively charged LDL subfraction by ion exchange HPLC. *Free Radic. Biol. Med.* 1991;11(3):247-53.

[112] Schauer R, Srinivasan GV, Wipfler D, Kniep B, Schwartz-Albiez R. O-Acetylated sialic acids and their role in immune defense. *Adv. Exp. Med. Biol.* 2011;705:525-48.

[113] Monti E, Bonten E, D'Azzo A, Bresciani R, Venerando B, Borsani G, Schauer R, Tettamanti G. Sialidases in vertebrates: a family of enzymes tailored for several cell functions. *Adv. Carbohydr. Chem. Biochem.* 2010;64:403-79.

[114] Sobenin IA, Salonen JT, Zhelankin AV, Melnichenko AA, Kaikkonen J, Bobryshev YV, Orekhov AN. Low density lipoprotein-containing circulating immune complexes: role in atherosclerosis and diagnostic value. *Biomed. Res. Int.* 2014;2014:205697.

[115] Sobenin IA, Karagodin VP, Melnichenko AC, Bobryshev YV, Orekhov AN. Diagnostic and prognostic value of low density lipoprotein-containing circulating immune complexes in atherosclerosis. *J. Clin. Immunol.* 2013;33(2):489-95.

[116] Tertov VV, Sobenin IA, Orekhov AN, Jaakkola O, Solakivi T, Nikkari T. Characteristics of low density lipoprotein isolated from circulating immune complexes. *Atherosclerosis.* 1996;122(2):191-9.

[117] Kacharava AG, Tertov VV, Orekhov AN. Autoantibodies against low-density lipoprotein and atherogenic potential of blood. *Ann Med.* 1993;25(6):551-5.

[118] Orekhov AN, Kalenich OS, Tertov VV, Novikov ID. Lipoprotein immune complexes as markers of atherosclerosis. *Int. J. Tissue React.* 1991;13(5):233-6.

[119] Schwaeble W, Schäfer MK, Petry F, Fink T, Knebel D, Weihe E, Loos M. Follicular dendritic cells, interdigitating cells, and cells of the monocyte-macrophage lineage are the C1q-producing sources in the spleen. Identification of specific cell types by in situ hybridization and immunohistochemical analysis. *J. Immunol.* 1995;155(10):4971-8.

[120] Bobryshev YV, Lord RS. Ultrastructural recognition of cells with dendritic cell morphology in human aortic intima. Contacting interactions of Vascular Dendritic Cells in athero-resistant and athero-prone areas of the normal aorta. *Arch. Histol. Cytol.* 1995;58(3):307-22.

[121] Bobryshev YV. Dendritic cells in atherosclerosis: current status of the problem and clinical relevance. *Eur. Heart J.* 2005;26(17):1700-4.

[122] Cao W, Bobryshev YV, Lord RS, Oakley RE, Lee SH, Lu J. Dendritic cells in the arterial wall express C1q: potential significance in atherogenesis. *Cardiovasc. Res.* 2003;60(1):175-86.

[123] Chistiakov DA, Sobenin IA, Orekhov AN, Bobryshev YV. Role of endoplasmic reticulum stress in atherosclerosis and diabetic macrovascular complications. *Biomed. Res. Int.* 2014;2014:610140.

[124] Orekhov AN, Tertov VV, Mukhin DN, Koteliansky VE, Glukhova MA, Khashimov KA, Smirnov VN. Association of low-density lipoprotein with particulate connective tissue matrix components enhances cholesterol accumulation in cultured subendothelial cells of human aorta. *Biochim. Biophys. Acta.* 1987;928(3):251-8.

[125] Orekhov AN, Tertov VV, Mukhin DN, Koteliansky VE, Glukhova MA, Frid MG, Sukhova GK, Khashimov KA, Smirnov VN. Insolubilization of low density lipoprotein induces cholesterol accumulation in cultured subendothelial cells of human aorta. *Atherosclerosis.* 1989;79(1):59-70.

In: Blood Lipids and Lipoproteins
Editor: Melissa R. Ruiz

ISBN: 978-1-63482-591-7
© 2015 Nova Science Publishers, Inc.

*Chapter 3*

# REGULATION OF THE INTAKE OF ARACHIDONIC ACID BY MODIFYING ANIMAL PRODUCTS AND ITS EFFECT ON INFLAMMATORY PROCESSES IN THE HUMAN BODY

*Dorota Bederska-Łojewska, Marek Pieszka, Paulina Szczurek, Sylwia Orczewska-Dudek and Mariusz Pietras*
Department of Animal Nutrition and Feed Science,
National Research Institute of Animal
Production, Kraków, Poland

## ABSTRACT

In this study we focused on the problem of arachidonic acid metabolism, belonging to the group of *n-6*. In recent years, there has been a marked increase in human consumption of polyunsaturated fatty acids *n-6*, with a simultaneous reduction of *n-3* intake. Arachidonic acid, a component of the lipid bilayer, is a precursor of many biologically important compounds include eicosanoids (prostaglandins, leukotrienes) involved in the stimulation of inflammatory processes. To prevent this, our diet should contain foods rich in *n-3*, which compete with *n-6* for the same metabolic pathways, thereby reducing the level of arachidonic acid in cells and extinguish the inflammation. This can be achieved by changing diet and modifying the composition of products of animal origin, such as eggs, milk and meat. Studies on the properties of *n-3* and *n-6* have been widely carried out in terms of their pharmacological use in the treatment of diseases with acute and chronic inflammation.

**Keywords:** Arachidonic acid, inflammatory process, omega acids, eicosanoids, metabolism of arachidonic acid

# ABBREVIATION

AA – arachidonic acid;
ALA – α-linolenic fatty acid;
CLA – conjugated linoleic acid;
COX – cyclooxygenase;
DGLA - dihomo-γ-linolenic acid;
DHA – docosahexaenoic acid;
DPA – docosapentaenoic acid;
EFAs – essential fatty acids;
EPA – eicozopentaenoic acid;
FFAs – unesterified fatty acids;
GLA – γ-linolenic acid;
LA – linoleic acid;
LCFAs – long chain fatty acids;
LCPUFAs – long chain polyunsaturated fatty acids;
LOX – lipoxygenase;
LT – leukotrienes;
NSAIDs – nonsteroidal anti-inflammatory drugs;
PG – prostaglandin;
PGI – prostacyclin;
PGHS – prostaglandin synthase;
PUFAs – polyunsaturated fatty acids;
TX – thromboxanes;
5-HETE – 5-Hydroxyeicosatetraenoic acid;
5-HPETE – 5-hydroperoxyeicosatetraenoic acid,
VEGF – vascular endothelial growth factor

# 1. INTRODUCTION

Long-chain polyunsaturated fatty acids (PUFAs) contain 18–20 carbons or more and can be grouped into two main families - *n-6* and *n-3*, depending on the position of the first double bond from the methyl end of the fatty acid (Venegas-Calerón et al., 2010). Arachidonic acid (AA) is a fatty acid with 20 carbon chain belonging to group *n-6*, what means its first double bond is located at the sixth carbon from the omega end.

AA next to docosahexaenoic acid (C22:6 DHA) is the major component of cell membranes. It is esterified in the membrane phospholipids at the sn-2 position and it occurs in abundant quantities in several tissues. The content of AA in plasma phospholipids and triglycerides is 8% and 1.64% respectively, when DHA content accounts for only 2.4% and 0.35% (Spector, 2000). In addition to the acids mentioned above also the linoleic acid (LA) from the *n-6* group (a precursor of arachidonic acid) constitutes a high lipid fraction (21%-23% of total fatty acids).

A large amount of PUFAs, especially AA, are presented in brain as well as in kidneys, heart, erythrocytes, neutrophils, monocytes and liver cells (Simopoulos, 2001; Wainwright, 2002; Palmquist, 2009). These fatty acids regulate several processes within the brain, such as neurotransmission, cell survival, neuroinflammation, mood and cognition. Overproduction or imbalance of the AA, DHA and their metabolites as well as their signaling pathways are impaired in various neurological disorders, including Alzheimer's disease and major depression (Kiso, 2011). AA is also a main fatty acid presented in placenta (Khan et al., 2008; Kremmyda et al., 2011). AA derived prostaglandins participate in the maintenance of pregnancy and initiation of labor.

Arachidonic acid plays an important role in inflammatory process and inflammation related diseases. While moderate inflammation may have positive effects on human and animal health, the chronic form contributes to a number of disorders. Increased consumption of AA leads to enhanced production of pro-inflammatory factors, thus intensifying the inflammation. Recent studies have demonstrated that AA might be also involved in pathogenesis of diseases correlated with central nervous system, cardiovascular system, diabetes and cancer (Maekawa et al., 2009; Russo, 2009; Vainio et al., 2011). More than 25% of today available for sale drugs were developed to target signaling pathways involving AA (Li et al., 2011).

## 2. BIOSYNTHESIS OF ARACHIDONIC FATTY ACID

The linoleic fatty acid (C18:2 LA) from the *n-6* group as well as α-linolenic fatty acid (C18:3 ALA) from the *n-3* group are termed as „essential fatty acids" (EFAs) because they are necessary for proper growth, development and function but cannot be synthesized by mammalian cells (Le et al., 2009). This disability results from the lack of adequate enzymes Δ12- and Δ15-desaturase and therefore these acids must be obtained directly from the diet (Simopoulos, 2009). The term "essential" was first proposed by Burr and Burr during their studies on essential fatty acids deficiency in rats (Burr and Burr, 1930). It was noticed only LA and ALA among many other fatty acids provided with diet could reverse the symptoms of the disease. LA and ALA are precursors for the rest of fatty acids including highly important metabolites like arachidonic acid (AA) from the *n-6* group as well as docosahexaenoic acid (C22:6 DHA) and eicosapentaenoic acid (C20:5 EPA) from the *n-3* group which can be provided directly in the diet or synthesized in the body (Wall et al., 2010). The amount of arachidonic acid in organism varies depending on the type and amount of fatty acids provided in the diet and their ratios. It is worth noting that conversion of EFAs occurs more efficiently and more intensively in animal tissues compared to human (Palmquist, 2009). Because production of AA, EPA and DHA might be limited in some conditions like prematurity and growth periods, what requires exogenous supplementation, they can be considered as conditional EFAs (Le et al., 2009).

Enzyme Δ-6 desaturase converts linoleic acid (C18:2 *n-6*) to γ-linolenic acid (GLA C18:3 *n-6*) by inserting a double bond between the sixth and seventh carbon (Figure 1). The same enzyme is involved in the metabolism of α-linolenic acid (C18:3 *n-3*) into octadecatetraenoic acid (C18:4). Then, GLA is converted to dihomo-γ-linolenic acid (DGLA C20:3 *n-6*) by the action of elongase which elongates carbon chain. As a result of Δ-5 desaturase activity DGLA is next transformed to the arachidonic acid (C20:4 *n-6*) by addition of a double bond between the fifth and sixth carbons. Elongase and Δ-5 desaturase synthesize also eicosapentaenoic acid (EPA C20:5 *n-3*) from octadecatetraenoic acid. Arachidonic acid can be then converted to tetraeicosapentaenoic fatty acid (C24:5 *n-6*) and eicosapentaenoic acid is changed into tetradocosaheksaenoic fatty acid (C24:6 *n-3*). The end products of fatty acids biosynthesis are docosapentaenoic (DPA C22:5 *n-6*) and docosahexaenoic acids (DHA C22:6 *n-3*) respectively.

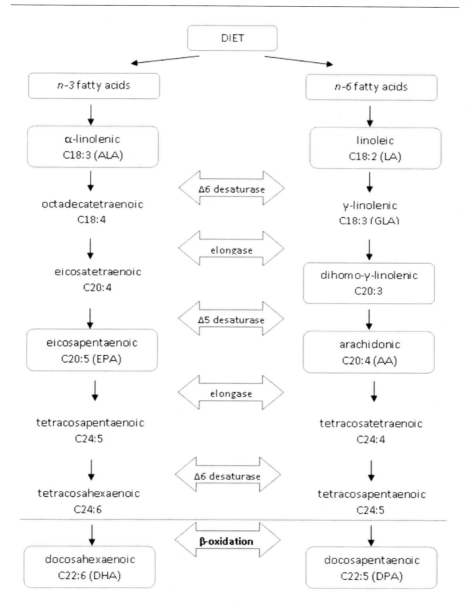

Figure 1. Biosynthesis of long chain polyunsaturated fatty acids from group *n-3* and *n-6*.

Desaturase and elongase are microsomal lipid layer associated enzymes which require the presence of zinc atoms for proper function (Nakamura et al., 2004). Transformation of ALA and LA depends mainly on the amount of desaturases, which activities decrease in the presence of saturated fats, cholesterol, trans-fatty acids, alcohol, adrenaline, insulin, glucocorticoids, diabetics and hypertensives (Das, 2008).

On the other hand factors such pyridoxine, zinc, nicotinic acid, and magnesium are co-factors for normal desaturase activity. The conversion efficiency of ALA-EPA-DHA varies also according to the sex. In adult men it is equal to 8% for reaction ALA-EPA and less than 0.1% for transformation of EPA to DHA, while the conversion efficiency to DHA in women is more than 9% (Abedi and Sahari, 2014). DHA concentration increase additionally during pregnancy and lactation due to estrogen action on Δ6-desaturase.

## 3. THE N-6/N-3 BALANCE

The amount of provided PUFAs it is not the only crucial factor for organism homeostasis and health. As it turned out, the ratio of *n-6/n-3* is even more significant as it is the key indicator of eicosanoids synthesis in the body. It follows from the fact that fatty acids from group *n-3* and *n-6* compete with each other for the same metabolic pathways. Thereby, long chain *n-3* PUFAs consumption extinguishes the inflammation by reducing the level of AA with simultaneous increase in EPA and DHA in cells due to reduced competition for Δ6-desaturase.

Unfortunately, enhanced marketing of cooking oils and margarines contributed to the increased intake of linoleic acid in Western countries over the last few years (Blasbalg et al., 2011). This changed pattern of consumption has caused a significant increase in the ratio of *n-6* to *n-3* PUFAs in the diet (Liu et al., 2015). This ratio should be approximately equal to 1, while nowadays in most Western populations it can be equal almost 20 (Burdge and Calder, 2006). The lowest ratio is observed among populations leaving in Coastal states, which consume a large quantity of fishes and seafoods rich in *n-3* fatty acids (Abedi and Sahari, 2014). A very good example are Japanese who take an ideal ratio of 1/2-4 (Aleksandra et al., 2009). The opposite site is represented by United States, where consumption of *n-6* is 10–30 times more than *n-3* (Abedi and Sahari, 2014). Not only a diet can influence the balance between *n-3* and *n-6* fatty acids. Some physiologic states such as aging, infant prematurity, hypertension and diabetes also can regulate long-chain *n-3* PUFAs production (Liu et al., 2015).

Clinical studies provide evidence that decreased $n-6/n-3$ ratio acts protective against degenerative disorders and cancer, as well as contributes to increased number of leukocytes, platelets and vascular endothelial growth factor (VEGF) (Russo, 2009). Moreover, diet changing in order to achieve a 4:1 ratio by replacing corn oil (high in LA) with olive and canola oil (low in LA) results in 70% decreasing in total mortality (de Lorgeril et al., 1994). Elevated ratio of AA to *n-3* fatty acid was also showed to be associated with depression (Lotricha et al., 2013).

## 4. ARACHIDONIC ACID BIOLOGY AND METABOLISM

Supplied with food AA, LA together with other polyunsaturated fatty acids are esterified in the liver and then released into the bloodstream, bounded to lipoproteins and distributed to all cells (Nowak, 2009). A small amounts of released AA, as well as other fatty acids are returned to the bloodstream where they bind to albumin (Figure 2). Over 90% of the fatty acids circulate in the blood bound to plasma proteins. The concentration of free fatty acids (FFAs) is relatively low but they are the most metabolically active lipid fractions. Plasma FFAs might be also a source of PUFA delivered to brain (Spector, 2000). Studies using radiolabeled PUFAs showed that they are entering brain readily when injected into the plasma with almost all AA cleared within 2 minutes (Rapoport et al., 2001; Liu et al., 2015).

FFAs participate not only in initiation of inflammatory pathways in a variety of cell types, but also produce numerous bioactive metabolites, function as secondary messengers and regulate platelet aggregation and vascular tone (Baker et al., 2011; Le et al., 2009). Moreover, they can directly affect the expression of genes which modulate fatty acids oxidation as well as fatty acids, lipids and lipoproteins synthesis (Kremmyda et al., 2011). Free fatty acids can stimulate the activity of different receptors and thus affect the secretion of insulin or neuropeptide Y (Doege and Stahl, 2006). The majority of FFAs is transported into the cell by diffusion but the part of them is moved by protein-mediated transport (McArthur et al., 1999). This includes transport of long chain fatty acids (LCFA) into the cells of muscle, liver, heart, adipose and intestines, where the metabolism and storage of LCFA occurre intensively (Doege and Stahl, 2006).

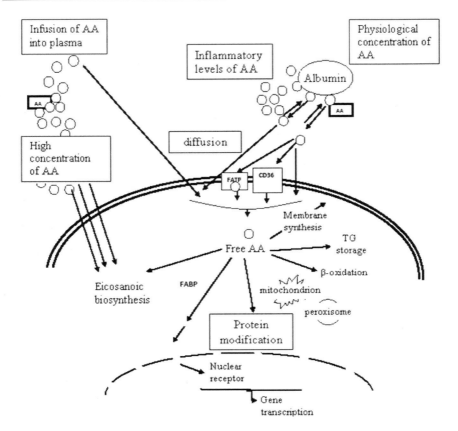

Figure 2. Metabolism of arachidonic fatty acid in the cells (adopted from Bederska-Łojewska et al., 2013).

Arachidonic acid may constitute a source of energy for cells trough β-oxidation process occurring in mitochondria as well as can be stored in form of energy reservoir in adipose tissue (Spector, 2000; Gao et al., 2009). When AA is released from membrane phospholipids it can either undergo resynthesis or be converted to eicosanoids (Doege and Stahl, 2006). Membrane proteins together with intracellular and extracellular receptors based on AA concentration decide which pathway will be processed. The concentration of AA in resting cell is kept at low level of 5 µM but the release of 1% of esterified AA from cell membranes may increase this level up to 50 µM (Brash, 2001). Introduction of high concentration of free AA is efficiently absorbed by the cells through diffusion and it is directly transformed to eicosanoids.

# 5. ARACHIDONIC ACID CASCADE AND INFLAMMATION

Among the many mediators of inflammation process being derivatives of polyunsaturated *n-3* and *n-6* acids (i.e. EPA - eicosapentaenoic acid, DHA - docosahexaenoic acid, DGLA - dihomo-γ-linolenic acid) those produced from arachidonic acid have the strongest biological activity (Harizi et al., 2008; James, 2010; Nowak, 2010). Inflammation is an important physiological phenomenon, without which it would be impossible to remove the pathological factor and repair the damages. Vasodilation of the blood vessels and their increased permeability enable the relevant proteins to migrate to damaged areas. The duration and severity of the inflammatory process depend on many factors such as location, size of the damage, personal profile of defensive reaction and the type of pathogen. Eicosanoids which include prostaglandins (PG), prostacyclin (PGI), thromboxanes (TX) and leukotrienes (LT) play a key role in the inflammatory process having an huge impact on the intensity and duration (Calder, 2006).

Typical inflammatory process last up to the moment when the pathological factor is removed and all damages are recovered in the body. However, in some cases, escalation of the immune response may appear which can be changed in the chronic form (Calder, 2006). Examples of diseases characterized by chronic inflammation are rheumatoid arthritis, inflammatory bowel disease (e.g. Crohn's disease), Type II diabetes, atherosclerosis, Alzheimer's disease, retinopathy, psoriasis, multiple sclerosis, chronic obstructive pulmonary disease, allergy and asthma (Calder, 2006; 2009; Nowak, 2010).

Eicosanoids are produced from arachidonic acid not directly but it requires a few reactions (Figure 3). In the first stage after the stimulation which may be: adrenaline, histamine, bradykinin, angiotensin II and thrombin, arachidonic acid is released from lipid bilayer by an enzyme phospholipase $A_2$. Then, with the involvement of cyclooxygenase (COX-1, COX-2), the molecules are converted to the cyclic eicosanoids: prostaglandins, prostacyclins and thromboxanes. Prostaglandin synthases: COX-1, COX-2 and PGHS exhibit their activities in the endoplasmic reticulum and lead to the transformation of arachidonic acid to $PGG_2$, reduced in the next step to $PGH_2$, which is a substrate for further synthesis of mentioned above eicosanoids.

Leukotrienes are also formed through multistep synthesis. As a result of an oxygen molecule insertion to the free arachidonic acid by 5-lipoxygenase, 5-hydroperoxyeicosatetraenoic acid (5-HPETE) is created. Then 5-HPETE can be reduced to two compounds: 5-hydroxyeicosatetraenoic acid (5-HETE, involving the glutathione peroxidase) or leukotriene $A_4$ (LTA$_4$, by further action of 5-lipoxygenase). LTA$_4$ is the precursor for the synthesis of other leukotrienes (Smith, 1989, Harizi et al., 2008). Furthermore, EPA can be also a substrate for lipoxygenase and COX, what gives a rise to eicosanoids with a slightly different structure to those formed from AA (Calder, 2009).

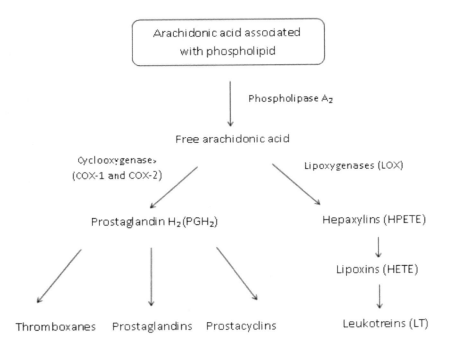

Figure 3. Scheme of the eicosanoids formation.

Eicosanoids exhibit a broad spectrum of action during the inflammatory response frequently performing contrary functions. The key for the proper conduction of the inflammatory process and for its extinction after deleting of all damages is correct interaction of these paracrine hormones. But sometimes, from different reasons, there is a need to mute the immune response. The inhibition of arachidonic acid cascade is a major mechanism of action of non-steroidal anti-inflammatory drugs.

## 5.1. Anti-Inflammatory Drugs

Non-steroidal and steroidal anti-inflammatory drugs have revolutionized the treatment of inflammatory diseases. The mechanism of action of the first one consists mainly on blocking eicosanoids synthesis via inhibition of cyclooxygenases (COX-1, COX-2). The first group of anti-inflammatory drugs (NSAIDs) include e.g. aspirin. At the heart of its action is inhibition of cyclooxygenases. By acetylation of serine 530 in the active site of enzymes the irreversible inhibition of both COX-1 and COX-2 occurs, which forbid the synthesis of eicosanoids. Aspirin demonstrates a much greater affinity for cyclooxygenase-1, what may contribute to the development of ulcers in the stomach or duodenum. Inhibition of prostaglandin synthesis in the gastrointestinal tract mainly through the inhibition of constitutive (permanently present in many cell types) COX-1 leads to a loss of protective effect of PG on gastric mucosa (Vane and Botting, 1998; Czyż and Watała, 2005). The mechanism of action of most non-steroidal anti-inflammatory, analgesic and antipyretic drugs i.e., indomethacin (acetic acid derivative), flubiprofen (ibufen derivative), naproxen (propionic acid derivative) or even ibufen rely on the inhibition of both cyclooxygenase-1 and cyclooxygenase-2 (Funk, 2001). However, in the case of ibufen there is no blocking of the active center, but competing with arachidonic acid for the substrate binding site of COX-1 and COX-2. The action of this drug is reversible and disappears when its decay (Thuresson et al., 2000). Improving the safety of anti-inflammatory compounds usage leads to the use of drugs selectively acting on COX-2 not constitutively secreted as COX-1 but only in response to inflammatory stimuli. Considered as very safe, but with a relatively weak action - paracetamol inhibits activity of the cyclooxygenase 3 (being a variant COX-1) present in the central nervous system – thereby there is observed decreased production of prostaglandins which results in a reduced sensitivity of the organism to the pain mediators (kinins, serotonin). However, due to the central action, there is no reduction in mucus secretion in the stomach and reduced barrier against hydrochloric acid. Importantly, paracetamol is characterized by ceiling effect, what means that the strength of its performance, above a certain dose, does not increase (Chandrasekharan et al., 2002).

In serious cases in the medicine there are used the narcotic drugs with analgesic action (morphine, codeine) and steroidal anti-inflammatory drugs. However, their main mechanism of action is not inhibition of arachidonic acid cascade. Narcotic drugs (opioids) act on specific receptors in the brain and spinal cord, which bind natural pain relief substances synthesized in the body (endorphins, encephalins) (Galligan et al., 1984; Schlaepfer et al., 1998). Steroidal anti-inflammatory drugs inhibit or stimulate expression of genes which protein products are involved in inflammatory processes. However, their use raises serious concerns, even though efficacy in the treatment of several autoimmune diseases, asthma or the use as immunosuppressive agents after transplantation is extremely high. Their long term administration carries a high risk of occurrence of adrenal insufficiency, increased susceptibility to bacterial, fungal and viral infections, diabetes, myopathy, glaucoma and osteoporosis (Świerczewska et al., 2013).

Each drug treatment is associated with a potential risk of side effects. A well-balanced diet can help the healing process. Regulation of intake of fats which are precursors of the inflammatory mediators can come to the rescue to people struggling with the disorder characterized by an excessive immune response.

## 6. ARACHIDONIC ACID AND BRAIN DEVELOPMENT

For decades, it is known that *n-3* fatty acids (eg. DHA) and *n-6* (eg. arachidonic acid) are extremely important in the normal development of the human brain in both prenatal and postnatal life. At the beginning, the growing fetus receives them, with other nutrients from the mother through the placenta, and in the postnatal period with natural milk or formula. At this time it is very important that the food contains a high concentration of polyunsaturated fatty acids. Proper supply of these acids in the diet is particularly important in humans starting from the fetal period up to the second year of life as at this time there is a rapid accumulation in the brain and its intensive development (Martinez, 1992). How important are the polyunsaturated fatty acids for the nervous system shows the fact that AA and DHA constitute 16% of the dry matter of the brain with their highest contents in gray matter (Pawełczyk et al., 2008).

The brain during development demonstrates the great need for arachidonic acid. AA in the body plays a very important role – it is the building block of cell membrane phospholipids in retinal photoreceptors and neurons in the brain (Karłowicz-Bodalska and Bodalski, 2007; Kolanowski, 2007). The proper functioning of neuronal membrane is a necessary condition for the undisturbed conduction of nerve impulses. At resting potential, the unequal distribution of cations on both sides of the cell membrane makes the interior of the cell more negative in comparison with the outer surface of the cell membrane.

Under the influence of stimulus, action potential appears and come to the rapid *influx* of sodium *ions* resulting in cell membrane depolarization. A moving depolarization wave along the cell membrane is a nerve impulse. When the signal reaches the end of the axon, the wave of depolarization goes to electrical or chemical synapse, where the signal is transmitted to the next cell. The cell which ended conductivity for some time remains in the lower excitability fase so-called hyperpolarisation, this prevent the wave of depolarization from going back. Potential changes in the lipid bilayer built from arachidonic acid are responsible for nerve cell activity (Michajlik and Ramotowski, 2003).

During the development of the nervous system arachidonic acid allows to create new neurons and later their maturation, the formation of synapse and correct brain plasticity. Numerous studies show that normal dietary intake of both AA and DHA promotes the maturation of brain structures important in cognitive processes such as the dorsolateral prefrontal cortex, supraorbital bark and associative parietal area (Willatts et al., 1998a; 1998b). Polyunsaturated fatty acids are largely responsible for specific structural and physicochemical properties of neuronal membranes. Their deficiency leads to disruption of neurotransmission through changes around fluidity of cell membranes, the tertiary structure of the membrane receptor and transport protein and the ligand-receptor interaction.

## 7. PRODUCTS OF ANIMAL ORIGIN AS A SOURCE OF ARACHIDONIC ACID AND ITS PRECURSORS

The type and amount of fat as well as proportion of fatty acids contained in the current diet have undergone major changes. Over the last 100 years there has been noticed a dramatic decrease in the consumption of polyunsaturated fatty acids from *n-3* group even up to 80%, and significantly increase of PUFAs from *n-6* group (Schwalfenberg, 2006; Russo, 2009). Such high intake of PUFAs *n-6* in the diet is caused by high consumption of vegetable oils (soybean oil, sunflower oil), due to their cholesterol-lowering properties in the serum (Simopolous, 2002; 2009). Excess of linoleic acid in the diet can cause adverse health effects (Margioris, 2009). The linoleic acid is metabolized to arachidonic acid which higher levels may contribute to prostaglandin and thromboxane overproduction (Margioris, 2009). Nowadays, the animal products such as meat and eggs are also a source of linoleic acid from PUFAs *n-6* (Russo, 2009), which means that it also comprises higher level of arachidonic acid and an increased load of AA for the consumer (Shin et al., 2012). In the human diet, the main source of AA are meat and eggs with smaller amounts coming from milk and fish (Huag et al., 2010). The content of LA and AA commonly found in different types of meat from animal housing under standard production system are summarized in Table 1.

**Table 1. Content of fat, arachidonic acid and linoleic acid in the lean meat of different species of livestock (Li et. al., 1998)**

| Type of meat | Fatty acids | | |
|---|---|---|---|
| | Total FAT [g/100g] | Arachidonic acid [mg/100g] | Linoleic acid [% of acids sum] |
| Beef | 1.4±0.2 | 28±5 | 5.6±1.1 |
| Lamb | 4.2±1.0 | 39±13 | 4.9±1.3 |
| Pork | 2.0±0.3 | 54±5 | 14.4±3.4 |
| Chicken | 2.3±0.5 | 43±10 | 14.4±1.7 |
| Duck | 1.7±0.3 | 99±38 | 18.2±3.4 |
| Turkey | 1.0±0.1 | 74±29 | 18.6±4.4 |

The fat content of lean meat ranges from 1g/100g in turkey meat to 4.2g/100g in lamb meat. Whereas the level of AA in beef and lamb meat is 1.5-2.5 times lower, compared to its content in meat of chicken, duck, turkey and pork respectively (Tab.1). Similar relationships refer to linoleic acid which is the precursor of AA, where there were 3 times lower levels of this acid in beef and lamb meat in comparison to chicken meat and pork. This is the result of different metabolism in ruminants: LA is degraded into monounsaturated and saturated fatty acids in the rumen by microbial biohydrogenation (70–95% and 85-100%, respectively) and only a small proportion, about 10% of dietary consumption, is available for incorporation into tissue lipids. Whereas, in monogastric animals the LA from diet is incorporated in unchanged form into tissue.

The content of fat as well as LA and AA is different in raw and cooked meat: beef, chicken, turkey, pork and also in eggs (Taber et al., 1998). It was also found that cooked meat and boiled eggs have shown higher content of AA and LA, up to 50-67% in comparison to the raw products. The average content of LA and AA in eggs is respectively: 1148mg/100g and 142mg/100g (Taber et al., 1998) but in milk the range is 2.0-7.0% and 0.03-0.09% (Chilliard et al., 2008).

# 8. POSSIBILITIES TO REDUCE THE CONTENT OF ARACHIDONIC ACID ON THE NUTRITIONAL WAY IN ANIMAL PRODUCTS

Increasing level of LA *n-6* is due to the production of compound feed, which the dominant components are grains (corn, barley, wheat and triticale), vegetable oils (sunflower, corn, soybean) and oilseeds characterized by a high content of *n-6* LA (Shin et al., 2012). The consequence of this situation is the increasing level of LA and AA *n-6* in meat lipids and the decreasing level of ALA *n-3*. The ratio of PUFAs *n-6/n-3* is particularly high in meat and depends largely on the composition of the ration used in the animals compound feed. It is estimated that this ratio can reached even 23:1, which is unfavorable from the point of view of consumer health. From the standpoint of human dietetics is preferable to limit the LA and AA in meat and meat products by enrichment animal product with PUFAs *n-3*.

The results of studies conducted in recent years have shown that the composition of fatty acids contained in the meat can be relatively easily modified by the diet (Barowicz and Pieszka, 2001; Kouba et al., 2003; Pietras and Orczewska-Dudek, 2013). It was found that PUFAs contained in the compound feed for monogastric animals are absorbed and incorporated in the unchanged form into tissue lipids which influences the fatty acid composition of the muscle (Flachowsky et al. 1997; Barowicz et al., 2002). A rich source of polyunsaturated fatty acids used in animal diet are oil seeds and oils from oilseeds. Numerous studies have demonstrated the beneficial effect of plant oil supplements on the fatty acid profile of lipids in meat and eggs (López-Ferrer et al., 1999, López-Ferrer et al., 2001; Barowicz and Pieszka, 2001; López-Bote et al., 2002; Kouba et al., 2003; Daza et al., 2005; Smink et al., 2010, Skiba et al., 2011; Pietras and Orczewska-Dudek, 2013) (Table 2). Similar modifications in fatty acid composition can be accomplished in the milk and eggs (Rego et al., 2005; Oliveira et al., 2010).

Recent studies of Haug et al. (2012) indicate that poultry meat as a potential source of *n-3* PUFAs in the diet of modern societies can contribute to reduce the risk of cardiovascular disease. The authors enriched the chicken meat with PUFAs *n-3*, using a mixture of linseed oil and rapeseed oil as a source of ALA in broiler chicken diet. The authors showed that enriched broiler chicken meat as a element of the human diet can increase the intake of EPA, DPA and DHA from ordinary food, but also decrease the intake of AA that is now too high. The similar results were obtained by Shin et al. (2012) when introducing flaxseed oil and long chain PUFAs (LCPUFAs) into chicken broiler diet. Also Coates et al. (2009) indicate that regular consumption of PUFAs *n-3* enriched pork can decrease the content of serum triglycerides and have beneficial effect on other biomarkers of cardiovascular health. They have also noticed that the AA levels in red blood cells in healthy subjects were diminished.

**Table 2. Effect of fodder or feed additives on reducing the level of linoleic and arachidonic acids in animal products**

| Fodder or feed additives | Animal specie/effect | Author/source |
|---|---|---|
| Linseed | Dairy cattle/↓C18:2; ↓C20:4 | Chilliard et al., 2008 |
| | Broiler chicken /↓C18:2; ↓C20:4 | Azcona et al., 2008 |
| Rape seed | Broiler chicken /↓C18:2; ↓C20:4 | Azcona et al., 2008 |
| Rape oil | Broiler chicken /↓C18:2; ↓C20:4 | Nobar et al., 2007 |
| Fish oil | Pigs/↓C18:2 | Skiba et al., 2011 |
| | Laying hens /↓C18:2; ↓C20:4 | Carvalho et al., 2009 |
| Linseed oil | Laying hens/↓C18:2; ↓C20:4 | Oliveria et al., 2010 |
| | Broiler chicken/↓C18:2; ↓C20:4 | Nobar et al., 2007 |
| | Broiler chicken/↓C18:2; ↓C20:4 | Shin et al., 2012 |
| Camelina oil | Broiler chicken /↓C18:2; ↓C20:4 | Pietras and Orczewska-Dudek, 2013 |
| | Laying hens /↓C18:2; ↓C20:4 | Cherian et al., 2009 |
| CLA | Pigs /↓C18:2 | Pieszka et al., 2006 |
| | Broiler chicken /↓C18:2 | Shin et al., 2011 |
| Pasture (grass) | Beef cattle /↓C18:2 | Noci et al., 2005 |
| | Lambs /↓C18:2 | Nuernberg et al., 2006 |
| Lard | Pigs/↓C18:2 | Skiba et al., 2011 |

The ruminant meat, especially lamb and the young cattle meat nourished on the basis of grazing system with a limited amount of compound feed characterized by favourable profile of PUFAs (Noci et al., 2005; Nuernberg et al., 2006). In beef cattle according to high biohydrogenation of unsaturated fatty acids, the use of fish oil is the most effective way of decreasing the AA level in meat together with increasing the level of *n-3* acids (Scollan et al., 2001). There were also trials on the protection of fatty acids from fish oil in feeding by microencapsulation conducted on dairy cattle what caused the significant increase of *n-3* acids level in milk (Lacasse et al., 2002). The easiest and cheapest method to protect the fatty acids in the rumen before biohydrogenation process and thus improving the utilization of the polyunsaturated fatty acids by ruminants, is introduced into the ration of whole oilseeds (Woods et al., 2008).

Feeding flax, camelina and marine oils has been also used to increase the *n-3* fatty acids content in poultry-based food products (Gonzales and Lesson, 2001; Pita et. al., 2010; Pietras and Orczewska-Dudek, 2013). Eggs from laying hens fed a diet containing graded levels of microencapsulated fish oil were characterized by lower content of AA and LA and grater content of EPA and DHA (Lawlor et al., 2010). Furthermore, the level of these LCPUFAs increased linearly with increasing level of microencapsulated fish oil in the hens diet.

Another way of lowering the levels of arachidonic and linoleic acids in the diet is the use of conjugated linoleic acid (CLA) isomers additives, which stimulate the activity of $\Delta 9$-, $\Delta 5$- and $\Delta 6$-desaturase (Eder et al., 2002, Smith et al., 2002, Shin et al., 2012). The use of CLA in compound feeds for pigs and broiler chickens increased levels of fatty acids of *n-3* group and decreased levels of *n-6* acids (Pieszka et al., 2006, Shin et al., 2011). Scientists also tried to change the fatty acid profile using minerals such as copper and chromium - addition of copper to feed of growing pigs resulted in an increase of content of saturated acids together with reduced level of unsaturated fatty acid in blood (Dove and Haydon, 1992), but an oversupply of these elements creates toxicological concerns and that is why there is no practical application in nutrition.

# CONCLUSION

The modern human diet is characterized by intake of very high levels of *n-6* fatty acids, what causes adverse for health increased ratio of acids *n-6/n-3* group (Asif, 2011). Recent studies show that excess of *n-6* fatty acids inhibit the metabolism of fatty acids *n-3*, what contributes to impaired physiological balance of synthesized from them biologically active compounds (Pike and Barlow, 2000; Marciniak-Łukasiak, 2011). Excessive intake of arachidonic acid and linoleic acid can be a cause of abnormal cell membrane permeability, blood coagulation and the overactive immune system, increase inflammation and may contribute to the development of neurodegenerative diseases (Asif,

2011). The proper ratio of both PUFAs groups is essential for maintaining homeostasis and may be extremely important in the treatment of diseases characterized by intensive inflammation. Overproduction of pro-inflammatory factors may be limited by consumption food containing large amount of *n-3* fatty acids. A properly balanced diet can help to obtain possibly the best profile of fatty acids contained in animal products and thus improve our health.

## REFERENCES

Abedi E. and Sahari M.A. (2014). Long-chain polyunsaturated fatty acid sources and evaluation of their nutritional and functional properties. *Food science & nutrition*, 2(5), 443-463.

Aleksandra A., Niveska P., Vesna V., Jasna T., Tamara P., Marija G. (2009). Milk in human nutrition: comparision of fatty acid profiles. *Acta. Vet.* 59:569–578.

Asif M. (2011). Health effect of omega-3,6,9 fatty acids: Perilla fructenses is a good example of plant oils. *Orient. Pharm. Exp. Med.*, 11: 51-59.

Azcona J.O., Schang M.J., Garcia P.T., Gallinger C., Ayerza R., Coates W. (2008). Omega-3 enriched broiler meat: The influence of dietary α-linolenic-ω-3 fatty acid sources on growth, performance and meat fatty acid composition. *Can. J. Anim. Sci.,* 88:257–269.

Baker R.G., Hayden M.S., Ghosh S. (2011). NF-κB, inflammation, and metabolic disease. *Cell metabolism*, 13(1), 11-22.

Barowicz T. and Pieszka M. (2001). Using linseed oil in fattening pig rations to modify chemical composition and dietetic value of pork. *Suppl. Polish J. Food Nutr. Sci.,* 3:42-45.

Barowicz T., Pieszka M., Pietras M., Migdał W., Kędzior W. (2002). Conjugated linoleic acid utilization for improvement of chemical composition and dietetic value of pork meat. *Ann. Anim. Sci.,* 2: 123-130.

Bederska-Łojewska D., Orczewska-Dudek S., Pieszka, M. (2013). Metabolism of arachidonic acid, its concentration in animal products and influence on inflammatory processes in the human body: a review. *Annals of Animal Science*, 13(2), 177-194.

Blasbalg T.L., Hibbeln J.R, Ramsden C.E., Majchrzak S.F., Rawlings R.R. (2011). Changes in consumption of omega-3 and omega-6 fatty acids in the United States during the 20th century. *Am. J. Clin. Nutr.*, 93, pp. 950–962

Brash A. (2001). Arachidonic acid as a bioactive molecue. *J. Clin. Inv.*, 11: 1339-1345.

Burdge G.C. and Calder P.C. (2006). Dietary α-linolenic acid and health-related outcomes: a metabolic perspective. *Nutr. Res. Rev.*, 19, pp. 26–52.

Burr G.O. and Burr M.M. (1930). On the nature and role of the fatty acids essential in nutrition. *J. Biol. Chem.*, p. 86.

Calder P.C. (2006). n-3 Polyunsaturated fatty acids, inflammation, and inflammatory diseases. *Am J Clin Nutr.*, 83:505–19.

Calder P.C.(2009). Polyunsaturated fatty acids and inflammatory processes: new twists in an old tale. *Biochimie.*, 91:791-795.

Carvalho P.R., Pita M,.C.G., Neto N.P., Mendonça C.X. (2009). Efficiency of PUFAs incorporation from marine source in yolk egg's laying hens. *Int. J. Poultry Sci.,* 8 (6): 603-614.

Chandrasekharan N.V., Dai H., Roos K.L., Evanson N.K., Tomsik J., Elton T.S., Simmons D.L. (2002). COX-3, a cyclooxygenase-1 variant inhibited by acetaminophen and other analgesic/antipyretic drugs: cloning, structure, and expression. *Proc. Natl. Acad. Sci. USA*; 99:13926–13931.

Cherian G., Campbell A., Parker T. (2009). Egg quality and lipid composition of eggs from hens fed Camelina sativa. *J. Appl. Poultry Res.*, 18: 143–150.

Chilliard Y., Martin C., Rouel J., Doreau M. (2008). Milk fatty acids in dairy cows fed whole crude linseed, extruded linseed, or linseed oil, and their relationship with methane output. *J. Dairy Sci.*, 92: 5199-5211.

Coates A.M., Sioutis S., Buckley J.D., Howe P.R. (2009). Regular consumption of n-3 fatty acid-enriched pork modifies cardiovascular risk factors. *Br J Nutr.,* 101(4):592-7.

Czyż M. and Watała C. (2005). Aspirin - wonderful panacea? Molecular mechanisms of action of acetylsalicylic acid in the body (in Polish). *Postępy Hig. Med. Dośw.*, 59: 105-115.

Das U.N. (2008). Can essential fatty acids reduce the burden of disease (s). *Lipids Health Dis*, 7(9).

Daza A., Rey A.I., Ruiz J., Lopez-Bote C.J. (2005). Effect feeding in free-range conditions or in confinement with different dietary MUFA/PUFA ratios and α-tocopheryl acetate, on antioxidants accumulation and oxidative stability in Iberian pigs. *Meat Sci.*, 69: 151-163.

Doege H. and Stahl A. (2006). Protein mediated fatty acid uptake: novel insights from in vivo models. *Physology*, 21: 259-268.

Dove C.R. and Haydon K.D. (1992). The effect of cooper and fat addition to the diets of weanling swine on growth performance and serum fatty acids. *J. Anim. Sci.*, 70: 805-811.

Eder K., Slomma N., Becker K. (2002). Trans-10, cis-12 Conjugated linoleic acid suppresses the desaturation of linoleic and α-linolenic acids in HepG2 cells, *J. Nutr.*, 132: 1115-1121.

Flachowsky G., Schöne F., Schaarmann G., Lübbe F., Böhme H. (1997). Influence of oilseeds in combination with vitamin E supplementation in the diet on backfat quality of pigs. *Anim. Feed Sci. Technol.*, 64: 91-100.

Funk C. (2001). Eicosanoid Biology Prostaglandins and Leukotrienes: Advances in eicosanoid biology. *Science*, 294; 1871-1875.

Galligan J.J., Mosberg H.I., Hurst R., Hruby V.J., Burks T.F. (1984). Cerebral delta opioid receptors mediate analgesia but not the intestinal motility effects of intracerebroventricularly administered opioids. *J Pharmacol Exp Ther.*, 229:641-648.

Gao F., Kiesewetter D., Chang L., Ma K., Bell J.M., Rapoport S., Igarashi M. (2009).Whole-body- synthesis-secretion rates of long-chain n-3 PUFAs from circulating unesterified α-linolenic acid in unanesthetised rats. *J. Lipid Res.*, 50: 750-758.

Harizi H., Corcuff J., Gualde N. (2008) Arachidonic-acid-derived eicosanoids: roles in biology and immunopathology. *Trends Mol Med.*, 10; 461-469.

Haug A., Nyquist N., Mosti T.J., Andersen M., Høstmark A.T. (2012). Increased EPA levels in serym phospholipids of humans after four weeks daily ingestion of one portion chicken fed linseed and rapessed oil. *Lipids in Health and Disease* 11: 104.

James J., Gibson R., Cleland L. (2000). Dietary polyunsaturated fatty acids and inflammatory mediator production. *Am J Clin Nutr*, 71; 343S–8S.

Karłowicz-Bodalska K. and Bodalski T. (2007). Unsaturated fatty acids, their biological and therapeutic importance (in Polish). *Postępy Fitoterapii*, 1; 46-56.

Khan A.H., Carson R.J., Nelson S.M. (2008). Prostaglandins in labor--a translational approach. *Front Biosci*; 13:5794-809.

Kolanowski W. (2007). Long-chain fatty acids omega-3 – the health importance in reducing the risk of lifestyle diseases (in Polish). *Bromat. Chem. Toksykol.*, 3: 229-237.

Kouba M., Enser M., Whittington F.M., Nute G.R., Wood J.D. (2003). Effect of a high-linolenic acid diet on lipogenic enzyme activites, fatty acid composition, and meat quality in the growing pig. *J. Anim. Sci.*, 81: 1967-1979.

Kremmyda L.S., Tvrzicka E., Stankova B., Zak A. (2011). Fatty acids as biocompounds: Their role in human metabolism, health and disease-a review. Part 2: Fatty acid physiological roles and applications in human health and disease. *Biomedical Papers*, 155(3), 195-218.

Lacasse P., Kennelly J.J., Delbercchi L., Ahnadi C.E. (2002). Addition of protected and unprotected fish oil to diets for dairy cows. I. Effects on the yield, composition and taste milk. *J. Dairy Sci.*, 69 (4), 511-520.

Lawlor J.B., Gaudette N., Dickson T., House J.D. (2010). Fatty acid profile and sensory char-acteristics of table-eggs from laying hens fed diets containing microencapsulated fish oil. *Anim. Feed Sci. Tech.*, 156: 97–103.

Le H.D., Meisel J.A., de Meijer V.E., Gura K.M. Puder M. (2009). The essentiality of arachidonic acid and docosahexaenoic acid. Prostaglandins, *Leukotrienes and Essential Fatty Acids*, 81(2), 165-170.

Li D., A. Ng, N.J. Mann, A.J. Sinclair (1998). Contribution of meat fat to dietary arachidonic acid. *Lipids*, 33 (4): 437-440.

Liu J.J., Green P., Mann J.J., Rapoport S.I. and Sublette M. E. (2015). Pathways of polyunsaturated fatty acid utilization: Implications for brain function in neuropsychiatric health and disease. *Brain research*, 1597, 220-246.

Lotricha, F.E., Searsb B., McNamarac R.K. (2013). Elevated ratio of arachidonic acid to long-chain omega-3 fatty acids predicts depression development following interferon-alpha treatment: Relationship with interleukin-6. *Brain, Behavior, and Immunity*, 31, 48–53.

López-Bote C.J., Isabel B., Daza A. (2002). Partial replacement of poly- with monounsaturated fatty acids and vitamin E supplementation in pigs diets: effect on fatty acid composition of subcutaneous and intramuscular fat and on fat and lean firmness. *Animal Sci.*, 75: 349-358.

Lopez-Ferrer S., Baucells M.D., Barroeta A.C., Galobart J., Grashorn M.A. (2001). n-3 enrichment of chicken meat. 2. Use of precursors of long-chain polyunsaturated fatty acids: linseed oils. *Poultry Sci.*, 80: 753–761.

Maekawa M., Takashima N., Matsumata M., Ikegami S., Kontani M., Hara Y. et al. (2009). Arachidonic acid drives postnatal neurogenesis and elicits a beneficial effect on prepulse inhibition, a biological trait of psychiatric illnesses. *PLoS One*, 4(4), e5085.

Marciniak-Łukasiak K. (2011). The role and importance of omega-3 (in Polish). *Żywność. Nauka.* Technologia. Jakość, 6 (79): 24-35.

Margioris AN. (2009). Fatty acids and postprandial inflammation. *Curr Opin Clin Nutr Metab Care.*, 12:129–37.

Martinez M. (1992). Tissue levels of polyunsaturated fatty acids during early human development, *J. Pediatr.*, 120; 129–S138.

McArthur M.J., Atshaves P.B., Frolov A., Foxworth W.D., Kier A.B., Schroeder F. (1999). Cellular uptake and intracellular trafficking of long chain fatty acid. *J. Lipid Res.*, 40: 1371-1383.

Michalik A. and Ramotowski W. (2003). Human Anatomy and Physiology (in Polish). Wydawnictwo lekarskie PZWL. Wydanie V. s. 461-466.

Nakamura T.M., Nara T.Y. (2004). Structure, function, and dietary regulation of Δ6, Δ5, and Δ9 desaturates. *Ann. Rev. Nutr.*, 24: 345-376.

Nobar R.S.D., Nazeradl K., Gorbani A. (2007). Effect of canola oil on saturated fatty acids contents in broiler meats. *J. Anim. Vet. Advances*, 6: 1204-1208.

Noci F., Monahan F.J., French P., Moloney A.P. (2005). The fatty acid composition of muscle fat and subcutaneous adipose tissue of pasture-fed beef heifers: Influence of the duration of grazing. *J. Anim. Sci.*, 83: 1167-1178.

Nowak J. (2010). Anti-inflammatory pro-resolving derivatives of omega-3 and omega-6 polyunsaturated fatty acids. *Postepy Hig. Med. Dosw.*, 64; 115-132.

Nowak J. (2009). Polyunsaturated fatty acids omega-3: biochemical, functional and practical aspects (in Polish). *Farmakoterapia w Psychiatrii i Neurologii*, 3-4: 127-146.

Nuernberg K., Ender K., Dannenberger D. (2006). Possibilities to produce healthy, tasty meat and to improve its nutritional value. *Pol. J. Food Nutr. Sci.*, 15/56 (1): 17-21.

Oliveira D.D., Baião N.C., Cancado S.V., Grinaldi R., Souza M.R., Lara L.J.C., Lana A.M.Q. (2010). Effects of lipid sources in the diet of laying hens on the fatty acid profiles of egg yolks. *Poultry Sci.*, 89: 2484–2490.

Palmquist D.L. (2009). Omega-3 Fatty Acid in Metabolism, Heath and Nutrition and for Modifies Animal Products Foods. *Prof. Anim. Sci.*, 25: 207-249.

Pawełczyk A., Pawełczyk T., Rabe-Jabłońska J. (2008). Exogenous polyunsaturated fatty acids may improve efficiency of selected cognitive functions. PSYCHiatr. *PSYCHol. KliN.*, 8;. 178-191.

Pieszka M., Paściak P., Janik A., Barowicz T., Wojtysiak D., Migdał W. (2006). The effect of sex and dietary antioxidants β-carotene, vitamins C and E in CLA-enriched diet on lipid profile and oxidative stability of pork meat. *J. Anim. Feed Sci.*, 15: 37–45.

Pietras M.P. and Orczewska-Dudek S. (2013). The effect of dietary Camelina sativa oil on quality of broiler chicken meat. *Annals Anim. Sci.*, 13(4): 1642-3402.

Pita M.C.G., Carvalho P.R., Piber Neto E., Mendonça Junior C.X. (2010). Effect of marine and vegetal sources on the hen diets on the PUFAs and PUFAs n-3 in laying hens egg yolk and plasm. *Int. J. Poultry Sci.*, 9: 148–151

Pike I.H. and Barlow S.M. (2000). The fats of life – the role of Fish. *Lipid. Technol.*, 12: 58-60.

Rapoport, S. I., Chang, M. C., & Spector, A. A. (2001). Delivery and turnover of plasma-derived essential PUFAs in mammalian brain. *Journal of lipid research*, 42(5), 678-685.

Rego O. A., Rosa H. J. D., Portugal P. V., Franco T., Vouzela C.M., Borba A.E.S., Bessa R.J.B. (2005). The effects of supplementation with sunflower and soybean oils on the fatty acid profile of milk fat from grazing dairy cows. *Anim. Res.* 54:17–24.

Russo G.L. (2009). Dietary n-6 and n-3 polyunsaturated fatty acids: From biochemistry to clinical implications in cardiovascular prevention. *Biochemical Pharmacology*, 77: 937-946.

Schlaepfer T.E., Strain E.C., Greenberg B.D., Preston K.L., Lancaster E., Bigelow G.E., Barta P.E., Pearlson G.D. (1998). Site of opioid action in the human brain: mu and kappa agonists' subjective and cerebral blood flow effects. *Am J Psychiatry*, 155; 470-473.

Scollan N.D., Choi N.J., Kurt E., Fisher A.V., Enser M., Wood J.D. (2001). Manipulating the fatty acid composition of muscle and adipose tissue in beef cattle. *Brit. J. Nutr.*, 85 (1): 115-124.

Shin D., Choi S.H., Go G., Park J.H., Narciso-Gaytan C., Morgan C.A., Smith S.B., Sanches-Plata M.X., Ruiz-Feria C.A. (2012). Effect of dietary combination of n-3 and n-9 fatty acids on the deposition of linoleic and arachidonic acid in broiler chicken meat. *Poultry Sci.*, 91: 1009-1017.

Shin D., Narciso-Gaytan C., Park J.H., Smith S.B., Sanches-Plata M.X., Ruiz-Feria C.A. (2011). Dietary combination effect of conjugated linoleic acid and flaxseed or fish oil on the concentration of linoleic and arachidonic acid in poultry meat. *Poultry Sci.*, 90, 1340–1347.

Simopoulos A (2001). Evolutionary aspects of diet and essential fatty acids. *World Rev. Nutr. Diet.*, 88: 18–27.

Simopoulos A.P. (2002). The importance of the ratio of omega-6/omega-3 essential fatty acids. *Biomed. Pharmacother.*, 56: 365–379.

Simopoulos, A. P. (2009). Omega-6/omega-3 essential fatty acids: biological effects. *World Rev Nutr Diet*, 99, 1-16.

Skiba G., Poławska E., Raj S., Weremko D., Czauderna M., Wojtasik M. (2011). The influence of dietary fatty acids on their metabolism in liver and subcutaneous fat in growing pigs. *J. Anim. Feed Sci.*, 20: 379–388.

Smink W., Gerrits W.J.J., Hovenier R., Geelen M.J.W., Verstegen M.W.A., Beynen A.C. (2010). Effect of dietary fat sources on fatty acid deposition and lipid metabolism in broiler chickens. *Poultry Sci.*, 89: 2432–2440.

Smith W. (1989). The eicosanoids and their biochemical mechanisms of action. *Biochem.*, 259; 315-324.

Smith S. B., Hively T. S., Cortese, G. M., Han, J. J., Chung K. Y., Casteñada P., Gilbert C. D., Adams V. L., Mersmann H. J. (2002). Conjugated linoleic acid depresses the delta-9 desaturase index and stearoyl coenzyme A desaturase enzyme activity in porcine subcutaneous adipose tissue. *J. Anim. Sci.*, 80:2110–2115.

Spector A.A. (2000). Plasma free fatty acid and lipoproteins as a source of polyunsaturated fatty acid for the brain. *J. Mol. Neurosci.*, 16: 159-165.

Świerczewska M., Ostalska-Nowicka D., Kempisty B., Nowicki M., Zabel M. (2013). Molecular basis of mechanisms of steroid resistance in children with nephritic syndrome. *Acta Biochim.* Pol. 60; 339-44.

Taber L., Chiu Ch-H., Whelan J. (1998). Assessment of the arachidonic acid content in foods commonly consumed in the American diet. *Lipids,* 33 (12): 1151–1157.

Thuresson E., Lakkides K., Smith W. (2000). Different Catalytically Competent Arrangements of Arachidonic Acid within the Cyclooxygenase Active Site of Prostaglandin Endoperoxide H Synthase-1 Lead to the Formation of Different Oxygenated Products. *J Biol Chem.*, 275; 8501-8507.

Vainio P., Gupta S., Ketola K., Mirtti T., Mpindi J. P., Kohonen P. et al. (2011). Arachidonic acid pathway members PLA2G7, HPGD, EPHX2, and CYP4F8 identified as putative novel therapeutic targets in prostate cancer. *The American journal of pathology*, 178(2), 525-536.

Vane J.R. and Botting R.M. (1998). Mechanism of action of nonsteroidal antiinflammatory drugs. *Am. J. Med.*, 104: 2–8.

Venegas-Calerón, M., Sayanova, O., & Napier, J. A. (2010). An alternative to fish oils: metabolic engineering of oil-seed crops to produce omega-3 long chain polyunsaturated fatty acids. *Progress in lipid research*, 49(2), 108-119.

Wainwright P.E. (2002). Dietary essential fatty acids and brain function: a developmental perspective on mechanisms. *Proc. Nutr. Soc.*, 61: 61-69.

Wall, R., Ross, R. P., Fitzgerald, G. F., & Stanton, C. (2010). Fatty acids from fish: the anti-inflammatory potential of long-chain omega-3 fatty acids. *Nutrition reviews*, 68(5), 280-289.

Willatts P., Forsyth J.S., DiModugno M.K. (1998a). Effect of long-chain polyunsaturated fatty acids in infant formula on problem solving at 10 months of age. *Lancet.*, 352: 688-691.

Willatts P., Forsyth J.S., DiModugno M.K. (1998b). Influence of long-chain polyunsaturated fatty acids on infant cognitive function. *Lipids*, 33; 973-980.

Wood J.D., Enser M., Fisher A.V., Nute G.R., Sheard P.R., Richardson R.I., Hug-hes S.I., Whittington F.M. (2008). Fat deposition, fatty acid composition and meat quality: *A review. Meat Sci.*, 78: 343-358.

# INDEX

## #

20th century, 72

## A

acetaminophen, 73
acetic acid, 65
acetylation, 65
acid, vii, viii, 17, 20, 25, 26, 27, 28, 29, 33, 34, 35, 36, 37, 51, 55, 56, 57, 58, 61, 62, 63, 64, 65, 66, 67, 68, 69, 71, 72, 73, 74, 75, 76, 77, 78, 79
action potential, 67
active site, 65
AD, 51, 52
additives, 71
adenosine, 2
adenosine triphosphate, 2
adipose, 61, 62, 76, 77, 78
adipose tissue, 62, 76, 77, 78
adrenal insufficiency, 66
adrenaline, 60, 63
adults, 8
aerobic exercise, 49
age, 2, 4, 5, 79
aggregation, 18, 47, 48, 49
albumin, 61
allergy, 63
alpha-tocopherol, 52

alters, 8
American Heart Association, 7
amino, 28, 32
amino groups, 28, 32
analgesic, 65, 73
angiotensin II, 63
antibody, 4, 39, 51
anti-inflammatory drugs, 56, 65
antioxidant, 46
antipyretic, 65, 73
aorta, 43, 46, 50, 53, 54
apoptosis, 2, 19, 46, 49
arachidonic acid, vii, viii, 55, 56, 57, 58, 63, 64, 65, 66, 67, 68, 70, 71, 72, 75, 77, 78
arterial cells, vii, 14, 19, 37
artery, 38, 47
assessment, 47
asthma, 63, 66
atherogenesis, 15, 19, 37, 53
atherosclerosis, vii, 9, 10, 13, 14, 15, 17, 19, 25, 26, 29, 37, 39, 40, 42, 43, 44, 50, 53, 54, 63
atherosclerotic plaque, 21, 23, 24, 38
atoms, 60
autoantibodies, 14, 15, 16, 19, 31, 32, 38, 39, 47
Autoantibodies, 30, 44, 53
autoimmune disease(s), 42, 66

## B

base, 47
BD, 10
beef, 68, 70, 76, 77
Belgium, 13
beneficial effect, 69, 70, 75
biochemistry, 77
biological activity, 63
biologically active compounds, 71
biomarkers, 70
biosynthesis, 58
blood, vii, 7, 8, 9, 10, 13, 14, 15, 16, 17, 19,
    31, 32, 33, 34, 36, 39, 41, 44, 50, 53, 61,
    63, 71
blood lipid levels, vii
blood plasma, vii, viii, 9, 14, 16, 17, 34, 39,
    41, 44
blood pressure, 8
blood vessels, 63
bloodstream, 17, 61
body fat, 8
bradykinin, 63
brain, 57, 61, 65, 66, 67, 75, 78
brain structure, 67

## C

cancer, 9, 57, 61
capillary, 18, 47
carbohydrate, vii, viii, 14, 20, 26, 27, 32, 41,
    46
carbon, 57, 58
carboxyl, 23
cardioprotective, vii, 1, 3
cardiovascular disease, 2, 3, 5, 7, 8, 9, 10,
    42, 43, 69
cardiovascular health, vii, 1, 2, 6, 70
cardiovascular risk, 3, 18, 45, 73
cardiovascular system, 57
carotene, 30, 76
carotenoids, 30
catabolism, 24
cattle, 70, 77

cell culture, 38
cell differentiation, 47
cell membranes, 57, 62, 67
cell surface, 21, 23
cell surface proteoglycans, 23
central nervous system, 57, 65
cerebral blood flow, 77
chemical, 29, 32, 33, 34, 44, 49, 67, 72
chemical properties, 34
chicken, 68, 69, 70, 74, 75, 76, 77
children, 78
cholesterol, vii, 1, 2, 4, 5, 6, 7, 8, 10, 15, 17,
    18, 19, 24, 26, 27, 29, 30, 34, 35, 37, 41,
    42, 43, 45, 46, 48, 51, 54, 60, 67
cholesteryl esters, vii, 13, 14, 19, 24, 41, 42,
    51
choline, 49
chromatography, 18, 20, 30, 33, 34, 36, 39,
    50
chromium, 71
chronic inflammation, viii, 56, 63
chronic obstructive pulmonary disease, 63
circulation, viii, 14, 15, 18, 41
cirrhosis, 42
City, 1
classes, 27
classification, 44
cloning, 73
clusters, 7
coenzyme, 30, 78
cognition, 57
cognitive function, 76, 79
cognitive process, 67
collagen, 25, 40
color, 28
competition, 60
complement, 7, 39
complications, 54
composition, viii, 3, 18, 26, 27, 33, 44, 50,
    51, 56, 69, 72, 73, 74, 75, 76, 77, 79
compounds, viii, 25, 29, 55, 63, 65
conduction, 64, 66
conductivity, 67
confinement, 73
conflict, 42

conflict of interest, 42
connective tissue, 25, 26, 38, 41, 54
consensus, 2, 4
consumption, viii, 55, 57, 60, 67, 68, 70, 72, 73
controlled trials, 8
cooking, 60
copper, 29, 71
coronary arteries, 43
coronary artery disease, 46, 51
coronary heart disease, 50
correlation(s), 5, 16, 30, 32
crops, 78
cultivation, 19
culture, 19, 25
culture medium, 25
CV, 47
cyclooxygenase, 56, 63, 65, 73
cytoplasm, 40

**D**

damages, 63, 64
decay, 65
deficiency, 58, 67
degradation, 21, 23, 34, 37, 38, 41, 42, 43
degradation rate, 23, 41
dendritic cell, 39, 53
depolarization, 67
deposition, vii, 13, 14, 41, 43, 77, 78, 79
depression, 61, 75
derivatives, 29, 63, 76
desialylated LDL, viii, 14, 16, 20, 21, 22, 23, 24, 25, 26, 27, 28, 29, 30, 32, 33, 34, 35, 37, 39, 41
detection, 36
diabetes, 4, 57, 61, 63, 66
diet, viii, 5, 55, 58, 60, 61, 66, 67, 68, 69, 71, 74, 76, 77, 78
dietary fat, 78
dietary intake, 67
diffusion, 61, 62
disability, 58
diseases, viii, 31, 56, 57, 63, 72, 74
disorder, 66

distribution, 27, 28, 44, 52, 66
DNA, 51
docosahexaenoic acid, 56, 57, 58, 63, 75
donors, 9, 37
dorsolateral prefrontal cortex, 67
drug treatment, 66
drugs, 3, 57, 65, 73, 78
dry matter, 66
duodenum, 65
dyslipidemia, 6

**E**

egg, 73, 76, 77
eicosapentaenoic acid, 58, 63
elastin, 25, 40
electronegative, viii, 14, 18, 28, 32, 33, 36, 39, 47, 48, 49, 52
electrophoresis, 16, 18, 27, 28, 45
ELISA, 4, 18, 47
elucidation, 36
endorphins, 66
endothelial cells, 18, 19, 39, 43, 47, 49, 50
energy, 62
engineering, 78
environment, 18
enzyme(s), 22, 36, 37, 50, 53, 58, 60, 63, 65, 74, 78
EPA, 56, 58, 60, 63, 64, 70, 71, 74
epitopes, 18, 32
erythrocytes, 57
essential fatty acids, 56, 58, 73, 77, 78
ester, 3, 15, 37, 41, 42, 43
estrogen, 60
evidence, 2, 61
excitability, 67
exercise, vii, 1, 2, 3, 4, 5, 6, 7, 8, 49
exposure, 18
extinction, 64
extracellular matrix, 25, 37, 41, 51

**F**

familial hypercholesterolemia, 52

families, 57
fasting, 9
fat, viii, 14, 15, 19, 30, 32, 41, 67, 68, 73, 75, 76, 77, 78
fatty acids, 27, 49, 56, 57, 58, 60, 61, 66, 67, 69, 70, 71, 72, 73, 74, 75, 76, 77, 79
feed additives, 70
fetus, 66
fibroblast growth factor, 46
fibroblasts, 11, 42
fibrosis, viii, 14
filters, 38
filtration, 36, 38
fish, 68, 70, 71, 75, 77, 78, 79
fish oil, 70, 71, 75, 77, 78
flotation, 16
foam cells, vii, 13, 14, 19, 39, 40
food, 61, 66, 70, 71, 72
food products, 71
formation, viii, 9, 14, 15, 17, 24, 37, 38, 41, 42, 44, 49, 64, 67
formula, 66, 79

## G

gastric mucosa, 65
gastrointestinal tract, 65
gel, 16, 18, 27, 28, 38, 45
genes, 9, 61, 66
genetic components, 6
genetic factors, 17
genome, 17
glaucoma, 66
glucocorticoids, 60
glucose, 10, 26
glutathione, 64
glycoproteins, 36, 37
glycosaminoglycans, 25
grass, 70
gray matter, 66
grazing, 70, 76, 77
growth, 17, 56, 58, 72, 73
growth factor, 56
guidelines, 7

## H

HDL, v, vii, 1, 2, 3, 4, 5, 6, 7, 8, 9, 10, 37
HDL quality, vii, 1, 3, 4, 6, 9
HDL-C, vii, 1, 2, 3, 4, 5, 6
healing, 66
health, vii, 1, 2, 6, 47, 57, 60, 67, 69, 70, 71, 73, 74, 75
health effects, 67
heart disease, 40
hepatocytes, 34
heterogeneity, 8, 9, 45
high density lipoprotein, 9, 45
high-density lipoprotein, vii, 1, 2, 4, 6, 7, 8, 9, 10
histamine, 63
HM, 44
homeostasis, 60, 72
hormones, 64
House, 75
housing, 68
human, vii, viii, 4, 9, 10, 13, 14, 19, 21, 22, 23, 24, 25, 31, 32, 33, 34, 36, 37, 38, 39, 40, 41, 42, 43, 44, 45, 46, 47, 48, 49, 50, 51, 52, 53, 54, 55, 57, 58, 66, 68, 69, 70, 71, 72, 74, 75, 77
human body, 72
human brain, 66, 77
human development, 75
human health, 75
human skin, 42
human subjects, 44, 52
hydrolysis, 21, 23, 24
hydroperoxides, 9, 29
hyperlipidemia, 51, 52
hypertension, 61
hypothesis, 29

## I

ID, 50, 53
ideal, 60
identification, 39
IL-8, 19

immune defense, 53
immune response, 19, 63, 64, 66
immune system, 71
immunoglobulin(s), 30, 39
immunosuppressive agent, 66
impulses, 66
in situ hybridization, 53
in vitro, viii, 9, 14, 15, 29, 32, 37, 38, 39, 41, 48, 51
in vivo, 9, 29, 30, 33, 38, 43, 46, 51, 73
Indians, 8
individuals, 4, 19, 26, 31, 34
induction, 49
infection, 9
inflammation, viii, 56, 57, 60, 63, 71, 72, 73, 75
inflammatory bowel disease, 63
inflammatory disease, 65, 73
inflammatory mediators, 66
ingestion, 74
inhibition, 2, 7, 64, 65, 75
inhibitor, 23, 38
initiation, 57, 61
insertion, 63
insulin, 47, 60, 61
insulin resistance, 47
interferon, 75
intervention, 4, 5, 6
intima, vii, 13, 14, 19, 21, 22, 23, 24, 25, 32, 34, 39, 40, 41, 43, 51, 53
ion-exchange, 18, 33
ionization, 18
ions, viii, 14, 15, 29, 37, 67
isolation, 18, 19, 29
isomers, 71
Italy, 33

**J**

Japan, 1

**K**

kidneys, 57

**L**

lactation, 60
LDL, vii, 2, 4, 13, 14, 15, 16, 17, 18, 19, 21, 22, 23, 24, 25, 26, 27, 28, 29, 30, 31, 32, 33, 34, 35, 36, 37, 38, 39, 40, 41, 43, 44, 45, 46, 47, 48, 49, 52
lead, 16, 19, 24, 29, 63
leisure, 8
leisure time, 8
lesions, viii, 14, 15, 16, 17, 19, 21, 24, 39, 47
leucine, 25
leukotrienes, viii, 55, 56, 63, 64
lifetime, 17, 37
ligand, 67
light, 17, 45
light scattering, 17, 45
linoleic acid, 56, 57, 58, 60, 67, 68, 71, 72, 74, 77, 78
lipid accumulation, viii, 14, 17, 19, 21, 25, 26, 28, 32, 37, 38, 39, 41, 42, 48, 50, 51
lipid metabolism, 78
lipid peroxidation, 29, 32, 43
lipid peroxides, 15
lipids, vii, 7, 8, 13, 14, 15, 19, 21, 24, 25, 33, 36, 40, 41, 50, 51, 61, 68, 69
lipoprotein particle, vii, viii, 14, 24, 29, 30, 32, 33, 34, 36, 37, 38, 41, 45
lipoproteins, vii, 2, 7, 8, 28, 29, 30, 31, 34, 36, 37, 41, 42, 43, 44, 46, 47, 48, 49, 50, 51, 52, 61, 78
liquid chromatography, 17, 29, 45
liver, 2, 17, 57, 61, 78
liver cells, 57
livestock, 68
locus, 45
low density lipoproteins, vii, 43, 46, 48, 49, 50, 51, 52
low-density lipoprotein, 2, 8, 14, 43, 46, 47, 48, 49, 50, 51, 52, 53, 54
lumen, 38
lycopene, 30
lysine, 18, 28, 32, 48

# M

macrophages, 2, 15, 22, 34, 39, 42, 43, 51
magnesium, 37, 60
magnetic resonance, 16
major depression, 57
majority, 19, 61
mammalian brain, 77
mammalian cells, 58
marketing, 60
mass, 4, 6
materials, 15
matrix, 25, 26, 38, 41, 54
MCP, 19
MCP-1, 19
measurement(s), 3, 4, 44, 47
meat, viii, 56, 68, 69, 70, 72, 74, 75, 76, 77, 79
median, 4, 6
medicine, 65
membrane permeability, 71
membranes, 67
messengers, 61
meta-analysis, 8, 44
Metabolic, 45
metabolic pathways, viii, 55, 60
metabolism, vii, viii, 2, 17, 50, 55, 56, 58, 61, 68, 71, 72, 74, 78
metabolites, 57, 58, 61
metabolized, 67
metals, viii, 14, 15
mitochondria, 62
models, 73
moderates, 2
modifications, 7, 16, 17, 28, 32, 33, 34, 36, 38, 41, 69
molecular weight, 52
molecules, 2, 19, 63
monoclonal antibody, 47
monosaccharide, 20
monounsaturated fatty acids, 75
morphine, 65
morphology, 53
mortality, 44, 61
Moscow, 13

MR, 8, 44
mucus, 65
multiple sclerosis, 63
myocardial infarction, 18, 47
myopathy, 66

# N

narcotic, 65
nephritic syndrome, 78
nerve, 66, 67
nervous system, 66, 67
neurodegenerative diseases, 71
neurogenesis, 75
neuroinflammation, 57
neurons, 66, 67
neurotransmission, 57, 67
neutral, 19, 27, 32, 34, 39
neutral lipids, 19, 27, 32, 39
neutrophils, 57
New South Wales, 13
nicotinic acid, 60
nitric oxide, 2
NMR, 18, 48
non-steroidal anti-inflammatory drugs, 64
normal development, 66
NSAIDs, 56, 65
nuclear magnetic resonance, 16, 29, 44, 45
nucleus, 40
nutrients, 66
nutrition, 71, 72, 73

# O

obesity, 45
oil, 61, 67, 69, 70, 71, 72, 73, 74, 76, 78
oleic acid, 24
omega-3, 72, 74, 75, 76, 77, 78, 79
opioids, 65, 74
organelles, 37
organism, 58, 60, 65
organs, 34
osteoporosis, 66
overproduction, 68

oxidation, 2, 4, 9, 14, 15, 17, 29, 30, 32, 34, 35, 36, 51, 61, 62
oxidation products, 15
oxidizability, 29, 30, 32, 48
oxygen, 63

**P**

pain, 65
parallel, 34
pasture, 76
pathogenesis, vii, 13, 14, 40, 57
pathology, 78
pathways, 61
permeability, 63
permission, 31, 35
peroxidation, 29, 36
pH, 23, 36
phagocytosis, 38
phenotype(s), 8, 45
phosphate, 27
phosphates, 29
phosphatidylcholine, 27
phosphatidylethanolamine, 27
phosphatidylserine, 27
phospholipids, 19, 28, 32, 39, 43, 57, 62, 66, 74
phosphorylation, 47
physicochemical properties, 67
pigs, 71, 73, 74, 75, 78
pilot study, 4, 5
placenta, 57, 66
plaque, 17
plasma proteins, 61
plasticity, 67
platelet aggregation, 61
platelets, 61
Poland, 55
polyacrylamide, 27
polysaccharide, 22
polysaccharide chains, 22
polyunsaturated fat, viii, 55, 56, 57, 59, 61, 66, 67, 69, 71, 72, 74, 75, 76, 77, 78, 79

polyunsaturated fatty acids, viii, 55, 56, 57, 59, 61, 66, 67, 69, 71, 74, 75, 76, 77, 78, 79
population, 4, 5
Portugal, 77
positive correlation, 30, 38, 39
poultry, 69, 71, 77
pregnancy, 57, 60
prematurity, 58, 61
preparation, 20, 30
prevention, 8, 44, 77
pro-atherogenic, 19
problem solving, 79
pro-inflammatory, 19, 57, 72
proliferation, viii, 14, 41
proline, 25
prostacyclins, 63
prostaglandins, viii, 55, 57, 63, 65
prostate cancer, 78
protection, 7, 70
protein structure, 18
proteinase, 23
proteins, 4, 9, 25, 31, 37, 62, 63
proteoglycans, 17, 19, 22, 23, 40, 41, 46
proteolysis, 23
proteolytic enzyme, 23
proteomics, 7
psoriasis, 63
psychiatric illness, 75
purification, 29, 46
pyridoxine, 60

**Q**

quantification, 45

**R**

RE, 45, 53
reactions, 63
reactive oxygen, 18
reactivity, 16
receptors, 18, 19, 41, 42, 51, 61, 62, 65, 74
recognition, 53

red blood cells, 70
relevance, 7, 49, 53
relief, 65
repair, 63
researchers, 17
residues, 3, 18, 48
resistance, 2, 52, 78
response, viii, 14, 15, 16, 32, 64, 65
resting potential, 66
reticulum, 40, 54, 63
retinopathy, 63
RH, 7
rheumatoid arthritis, 63
risk, 8, 9, 49, 66, 69, 74
Russia, 13, 41

**S**

safety, 65
sample design, 4, 5
saturated fat, 60, 68, 76
saturated fatty acids, 68, 76
scattering, 27, 45
science, 72
secretion, 43, 61, 65, 74
seed, 70, 78
sensitivity, 65
serine, 23, 65
serotonin, 65
serum, 10, 19, 25, 34, 36, 39, 40, 42, 44, 45, 47, 50, 51, 52, 67, 70, 73
sex, 60, 76
shape, 16
showing, 5, 6
sialic acid, 17, 20, 26, 27, 28, 32, 33, 34, 36, 37, 51, 53
side effects, 66
signaling pathway, 57
signalling, 18
smoking, 18
smooth muscle, 19, 22, 34, 41, 51
smooth muscle cells, 19, 22, 34, 41, 51
sodium, 29, 42, 67
SP, 45
species, 18, 68

spinal cord, 65
spleen, 39, 53
SS, 7, 43
stability, 49, 73, 76
standard deviation, 4, 6
standardization, 17
states, 60
statin, 44
sterols, 29
stimulation, viii, 37, 41, 51, 55, 63
stimulus, 67
stomach, 65
storage, 42, 61
stress, 40, 54
stroke, 7
structural changes, 15
structure, 18, 28, 39, 48, 64, 67, 73
style, 10
subgroups, 44
substrate, 37, 63, 64, 65
sulfate, 25
Sun, 9, 44
supplementation, 58, 74, 75, 77
surface properties, 18
survival, 57
susceptibility, 9, 10, 29, 34, 46, 52, 66
symptoms, 58
synapse, 67
synthesis, 25, 26, 37, 41, 51, 60, 61, 63, 65, 74

**T**

target, 9, 57
techniques, 16
therapeutic targets, 42, 78
therapy, 7
thrombin, 63
thrombosis, 2
thromboxanes, 56, 63
tissue, 50, 68, 69
tocopherols, 30
total cholesterol, 34
trafficking, 76
training, 8

transcription, 46
transferrin, 37
transformation, 40, 60, 63
transplantation, 66
transport, 2, 7, 10, 61, 67
treatment, viii, 4, 42, 56, 65, 66, 72, 75
triggers, 47
triglycerides, 19, 27, 28, 57, 70
trypsin, 23
tryptophan, 18
Turkey, 68
turnover, 77
type 2 diabetes, 8, 18, 44
tyrosine, 3

**U**

UK, 10
United States, 60, 72
USA, 73

**V**

vacuole, 40
vascular endothelial growth factor (VEGF),
    61
vascular wall, viii, 14, 15

VCAM, 19
vegetable oil, 67, 69
vertebrates, 53
viral infection, 66
vitamin E, 74, 75
vitamins, 17, 76
VLDL, 37

**W**

walking, 8
weight loss, 5
Western countries, 60
white blood cells, 34
workers, 15, 32

**Y**

yield, 75
yolk, 73, 77

**Z**

zinc, 60